Roads

EXPERTISE

**CULTURES AND
TECHNOLOGIES
OF KNOWLEDGE**

EDITED BY DOMINIC BOYER

A list of titles in this series is available at www.cornellpress.cornell.edu.

Roads

An Anthropology of Infrastructure and Expertise

Penny Harvey and Hannah Knox

Cornell University Press
Ithaca and London

First published 2015 by Cornell University Press
First printing, Cornell Paperbacks, 2015

Printed in the United States of America

Library of Congress Cataloging-in-Publication Data

Harvey, Penelope, 1956– author.
  Roads : an anthropology of infrastructure and expertise / Penny Harvey
and Hannah Knox.
      pages cm— (Expertise (Ithaca, N.Y.))
  Includes bibliographical references and index.
  ISBN 978-0-8014-5323-6 (cloth : alk. paper)
  ISBN 978-0-8014-7964-9 (pbk. : alk. paper)
  1. Infrastructure (Economics)—Political aspects—Peru.
2. Infrastructure (Economics)—Social aspects—Peru.   3. Roads—
Political aspects—Peru.   4. Roads—Social aspects—Peru.
5. Ethnology—Peru.   I. Knox, Hannah, 1977– author.   II. Title.
III. Series: Expertise (Ithaca, N.Y.).

  HC79.C3H368 2015
  388.10985—dc23
                    2014040304

Cloth printing   10 9 8 7 6 5 4 3 2 1
Paperback printing   10 9 8 7 6 5 4 3 2 1

The figure and all photos are by the authors.

Contents

PREFACE

We began to talk to each other about doing a joint ethnography of Peru's roads in 2004. Over the ensuing ten years we have accrued many debts to those who have variously inspired, assisted, cajoled, critiqued, signposted, reassured, and supported us in bringing the book to completion. Although it is now impossible to reconstruct the specifics of many of these conversations, they have all undoubtedly affected our thinking about what roads are, what they do, and what they can reveal about our contemporary world. Indeed, roads have proved to be a remarkably generative topic of conversation, inspiring comment, insight, and opinion from disparate family members, friends, and colleagues in the United Kingdom, Peru, and beyond. It has been an unexpected delight to find ourselves working on something that seems to make sense to people, particularly in Peru. Furthermore, in the world of social research more generally, "roads" seems to lend themselves to a huge range of intellectual concerns.

Without belittling the importance of all of these interactions, there are inevitably particular people, moments, and settings that stand out as having been central to the realization of the book that we want to acknowledge more directly. The acknowledgments also serve another purpose. Although necessarily somewhat curtailed, the tracing of a few key relationships allows us to produce an account of how we did the research. The book does not contain a chapter that explicitly outlines our methods. The ethnography emerged through the extensive and largely serendipitous discussions and encounters that gradually provoked and guided us into shaping this book in the way we have. To put it another way, to acknowledge the invaluable support that we have received from so many people since we began our research into the roads of Peru is to describe our method—the ethnographic method, which looks to learn through engagement, dialogue, and living alongside other people as they go about their daily lives, whether in Peru or in Manchester, at home or at conferences and workshops. Our point is that ethnographic research is a relational affair. We learn through the relationships in which we become entangled. These relationships have been the bedrock of our research, from the moment that we formed the idea of what to focus on, to when we began to track that idea through the encounters that became the empirical foundations of the work, through to the conversations and critical challenges that encouraged us to report back and to keep thinking about what we were learning.

It is hard to remember exactly how the topic of roads took on the momentum it did for us, and there are doubtless diverse origins. A workshop on anthropology and the state organized by Christian Krohn-Hansen and Knut Nustad in Oslo in 2002 provided the impetus to think about roads as an ethnographic way into the exploration of state effects. Many subsequent visits to Norway, both to Bergen and to Oslo, have continued to provide a consistent source of inspiration and support. Special thanks are due here to Bruce Kapferer for his intellectual companionship. Our interest in roads also emerged out of prior work that we had done on the anthropology of technology more broadly conceived. We had both worked ethnographically on the promise and challenges of new information and communications technologies, Penny through the UK Economic and Social Research Council's "Virtual Society?" program, and Hannah as a researcher in the Economic and Social Research Council's "Evolution of Business Knowledge" program. From early on we approached roads in terms of the role they play as an older form of communications technology, drawing on our thinking that had been developed through our prior collaborative research with Sarah Green, Damian O'Doherty, Theo Vurdubakis, and Chris Westrup.

However, it was under the auspices of the ESRC Centre for Research on Socio-Cultural Change (CRESC) that the conditions arose to turn an intriguing research possibility into a funded research project. For that we must thank the original directors of CRESC, Tony Bennett, Mike Savage, and Karel Williams, and their vision for a research center that would provide a rich intellectual space within which an open and exploratory empirical research project like this could be developed. The ESRC's Small Grants scheme provided a grant in 2005 to allow Penny to undertake a year's fieldwork in Peru in 2005–06. This small grant was lodged inside CRESC, where Hannah worked from 2004 to 2014. Penny was subsequently seconded to work at CRESC, as a director since 2006. CRESC thus provided the intellectual home for this project, allowing us to extend our very strong and important ties to the Department of Social Anthropology at the University of Manchester into new conversations.

Our home within CRESC was a research group that came to be known colloquially as "Theme 4." The group was initially concerned with approaching social change from the perspective of politics and cultural values, a topic that over time took shape as a conversation between anthropological and ethnographic approaches to social change, actor-network theory, Deleuzian-inspired poststructural philosophy, theories of affect, and new cultural materialisms. We have been fortunate to be surrounded in Theme 4 by our inspirational anthropology colleagues: Jeanette Edwards, Adolfo Estalella, Gillian Evans, Gemma John, Fabiana Li, Cecilia Odegaard, Madeleine Reeves, and Peter Wade. We also have been challenged in Theme 4 to account for our anthropological perspective by our intellectual friends from other disciplines: Elizabeth Silva, Nick Thoburn, and Kath Woodward from sociology; Eleanor Casella from archaeology; Christine McClean and Damian O'Doherty from management and organization studies; Tone Huse from geography; Michelle Bastian from philosophy; and Andy Gale from engineering. It was serendipity rather than strategy that brought together this particular mix of scholars, but the creation of this environment of curiosity, supportive critique, and theoretical creativity would never have been possible without the model of open intellectual enquiry that CRESC put in place.

Beyond Theme 4, we have also been fortunate to develop our thinking on the relationship between the seemingly diverse topics of road construction, political economy, materiality, and public works through our participation in the broader intellectual life of CRESC. In developing our thinking about the form that an anthropology of infrastructure might take, we have drawn particular inspiration from work on "the social life of method" by John Law, Niamh Moore, Mike Savage, and Evelyn Ruppert; from Francis Dodsworth and Sophie Watson's work on "city materialities"; from Tony Bennett and Patrick Joyce's exploration of "material power"; and from the work of Karel Williams and his colleagues on "financialization and the multiple logics of capitalism."

However, our work on roads would not have been possible without the insights and support from the many friends that Penny had made during her years of working in Peru. In Lima, the generosity and support of Cecilia Blondet was constant, and her neighborhood, her home, and her family opened up a whole world of debates and discussions on all things political, cultural, and even theological. The challenge of translating our ethnographic experiences on roads in ways that made sense to our friends in Lima who knew Peru so deeply, and yet so differently, allowed us a privileged insight into the complexities involved in confronting the inequalities of contemporary Peru. The companionship in these conversations of Natalia González, Carmen Montero, and Hortensia Muñoz was especially appreciated. Marisol de la Cadena was the other person from Lima who had a fundamental influence on the trajectory of our project. She was always ahead of the curve in recognizing how our work might engage intellectual agendas in anthropology and in science and technology studies, and in understanding how ethnographic insights could transform the ways in which we approach questions of power and social transformation. She never failed to inspire and to open new avenues of investigation. Aroma de la Cadena and Eloy Neira were also remarkably generous with their time and became key guides to where we should work, providing many practical suggestions, leads, and contacts that were grounded in their own deep and engaged understanding of contemporary Peru. It was Eloy who told us about the Iquitos-Nauta road, encouraging us to visit and giving us key contacts to get the project under way. Once we were focused on Iquitos-Nauta, Alberto Chirif, Federika Barclay, and Manuel Cornejo were also very important guides to the history and ethnography of this region of Peru.

In Cusco, where Penny had worked since 1983, other long-established relationships were fundamental to us. Matilde Villafuerte, Penny's first Quechua teacher, and her husband, Julio Cesar Guevarra, had always been like family to Penny and welcomed Hannah into the fold. Julio, a civil engineer, offered an important critical perspective and informed our early understandings of what it takes to build a road, introducing us to people such as Ing. Julio Bonino and Ing. Guido Vayro, who were able to talk about road construction in southern Peru from long experience.[1] In Ocongate (a town that in the 1980s had been a whole day's journey from Cusco), old friends were talking about the proposals for the new Interoceanic Highway. Long before the camps were built, Zoraida and Raúl Rosas were preparing accommodation for the first engineers sent to scope out the route. Zoraida and Raúl were always busy, and yet always had time to talk and to just let us be around, and their house remained open to us and provided a fortuitous space for our initial engagement with the engineering company. Pascuala Quispe, Claudio Machaka, and their children were our generous hosts for much of the time we stayed in the Ocongate area. They knew about construction work, contract labor, and the precarious conditions of life for rural agriculturalists. Friends with Penny for nearly thirty years, this family allowed us to reflect on technological and political change from the perspective of those who were often abstractly represented as the "beneficiaries" of road development projects. Their concerns and understandings of the construction process were crucial to our thinking.

Our contacts in Lima, Cusco, and Ocongate had helped us to articulate a project to study the roads of Peru, but it still remained for us to find those people who were responsible for constructing them. When Penny first visited Nauta she was introduced to Ing. Juan Gonzalez, the head of the company of construction engineers that was then engaged in completing the road that we describe in detail in the coming chapters. Juan gave us our first taste of what a road construction project involved. With an enthusiasm that we were to come across with great frequency among the various engineers that we got to know, he was soon offering us a guided tour of the works, driving some distance to point out all the things that they were doing, and introducing colleagues with instructions that they should show us around and explain how things worked. This is how we met Gulliver, a civil engineer who specialized in soil analysis. Gulliver not only tolerated our constant questions about the basics of the construction process on the Iquitos-Nauta road but allowed us to follow him to the Interoceanic Highway project, where he helped us find a tangible route into what was a hugely complex multinational construction project. He gave us our first view of the construction laboratories and introduced us to his colleagues Alberto, Enrique, Fabio, and Melissa, from whom we then found our way into other parts of the organization. Armed with a basic understanding of the organizational structure of the project that we garnered from Gulliver and his engineering colleagues in Puerto Maldonado, we began to pursue other points of entry into the consortium: conducting interviews with senior management in Lima, approaching the headquarters of the consortium in Cusco, and gradually making ourselves a known presence on the project. This involved some persistence on our part and a lot of sitting around waiting for lifts and meetings, but it soon paid off. After negotiating the initial hurdles of access, we found ourselves joining engineers on the construction sites, attending health and safety briefings, and shadowing the community relations team. With remarkable openness we were welcomed into the engineering camps, where we were accommodated and given a glimpse of day-to-day life in these large construction projects. We were constantly amazed by the openness and supportiveness of construction firms toward our project and by the assistance we received from many members of the CONIRSA team, including Roberto Campelo, Paulo Campos, Cristobal Corpancho, Alfredo Delgado, Richard Diaz, Silvana Leon, Rosana Lopez, Delsy Machado, Toribio Naula, José Talavera, Silvia Tejada, Hugo Vargas, and Alberto Wanderley Soares.[2]

Throughout our research we were concerned to hold in view not just the perspective of the engineers and their colleagues but also those of the residents of the towns through which the roads passed. For facilitating meeting with local residents, we thank Alejandro Diaz and his nephew Bernadino, Miguel Alfredo Flores, Waldemar Vergel Vela, and Padre Walker. And for insights into what a road construction project looks like from the perspective of local residents, we are particularly grateful to Pascual Aquituari, Brendy Sandoval, and Laura Conde Tamani for the time they spent with us on the Iquitos-Nauta road, and to Eufemia Quispe for her hospitality in Puerto Maldonado. In addition, our understanding of the histories of these two roads was enriched by a number of remarkable and unforgettable conversations with some of the older residents about their lives along these roads. We thank Segundino Puglia Cruz, Don Fidel of Nauta, Benedicto

Kalinowski and his family, José Domingo Murayari, Braulio Reina, Alejandro Uraco, and Victor Miranda Vicente for their generosity with their time and for sharing their experiences with us.

We would also like to acknowledge the insights and ethnographic perspicacity of Tom Grisaffi, who spent one summer living and traveling on the Interoceanic Highway, making notes and taking photographs on our behalf. His ability to make friends and to engage people allowed him to live alongside the young men working on the construction projects in ways that greatly extended the relationships that we had made. Furthermore, he provided valuable reflections on the specificities of Peru by comparison with Bolivia, where he had conducted his doctoral research. The companionship of Gonzalo Valderrama on two short research trips also opened doors and new perspectives.

In the several years that have passed since we completed the fieldwork, we have been drawn into various research initiatives that have influenced the ways in which we interrogated our ethnographic materials. Penny's research collaboration with Deborah Poole on regional government in Peru has dramatically enriched our understandings of the distributed and experimental processes that comprise what gets loosely referred to as the Peruvian state. This ethnographic research, conducted with Jimena Lynch Cisneros, Teresa Tupayachi Mar, and Annabel Pinker, accompanied the completion of this book—constantly inflecting our understandings and opening up new possibilities. During this period we have also had many opportunities to present our work at home and to international audiences, and we are grateful to all those who participated in these meetings and gave us feedback. For organizing events that we have found particularly inspirational, we thank Tony Bennett, Matt Candea, Jo Cook, Alberto Corsin-Jimenez, Laura Hubbard, Celia Lury, Evelyn Ruppert, Catherine Trundle, Christopher Vasantkumar, and Tom Yarrow. And for the specific collaborations on writing up the roads materials in a more comparative framework, we are also grateful to Dimitris Dalakoglou and to Soumhya Venkatesan.

Getting the manuscript to completion would not have been possible without the enthusiasm and support of those at Cornell University Press: Dominic Boyer, Peter Potter, Kitty Hue-Tsung Liu, and Max Porter Richman.

And last, but not least, we thank our families. The writing of this book has been punctuated by the arrival of Hannah's three daughters, Imogen,

Francesca, and Beatrice, who have provided both a healthy distraction from and justification for writing. Combining the care of three children with academic life would not have been possible without the ongoing intellectual engagement and practical support of Damian O'Doherty. Penny's daughter, Laurie, grew up with this book, having to put up with fieldwork absences and a mother who spent far too long on the computer throughout her teenage years. Laurie was constantly amazed that anybody could be dedicated to a topic so devoid of glamour. Ben Campbell did so much more than hold the fort, engaging and extending the intellectual work and making sure that we all had some fun and found joy along the way.

Short sections of this monograph have previously appeared in other publications. A few short passages of chapter 1 are duplicated in Penny Harvey and Hannah Knox, "'Otherwise Engaged': Culture, Deviance, and the Quest for Connectivity through Road Construction," *Journal of Cultural Economy* 1 (1) (2008): 79–92. Parts of chapter 4 appear in Hannah Knox and Penny Harvey, "Anticipating Harm: Regulation and Irregularity on a Road Construction Project in the Peruvian Andes," *Theory, Culture and Society* 28 (6) (2011): 142–63. Sections of chapter 7 are also reproduced in Penny Harvey and Hannah Knox, "Surface Dramas, Knowledge Gaps and Scalar Shifts: Infrastructural Engineering in Sacred Spaces," Occasional Paper Series 4, no. 2 (2013) (Penrith, New South Wales: Institute for Culture and Society, University of Western Sydney).

Abbreviations

AASHTO	American Association of State Highway and Transportation Officials
CAF	Corporación Andina de Fomento
CTAR-Loreto	Consejo Transitorio de Administración Regional de la Región Loreto
IFI	international financial institution
IIAP	Instituto de Investigaciones de la Amazonía Peruana
IIRSA	Initiative for the Integration of the Regional Infrastructure of South America
MRTA	Movimiento Revolucionario Túpac Amaru (Shining Path)
MTC	Ministerio de Transportes y Comunicaciones
NGO	nongovernmental organization
SINAMOS	Sistema Nacional de Apoyo a la Movilización Social
SNIP	Sistema Nacional de Inversión Pública
WWF	World Wildlife Fund

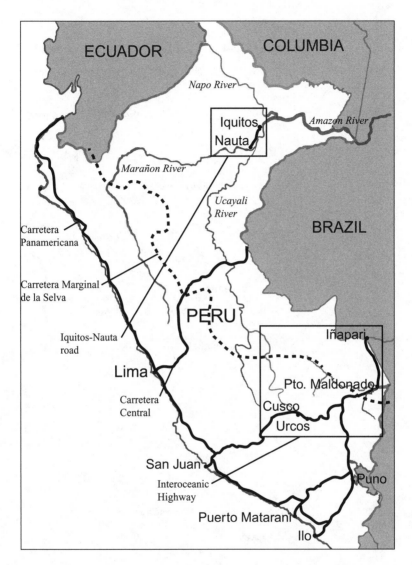

Map 1. Peru's major roads

Map 2. The Interoceanic Highway/Route 26

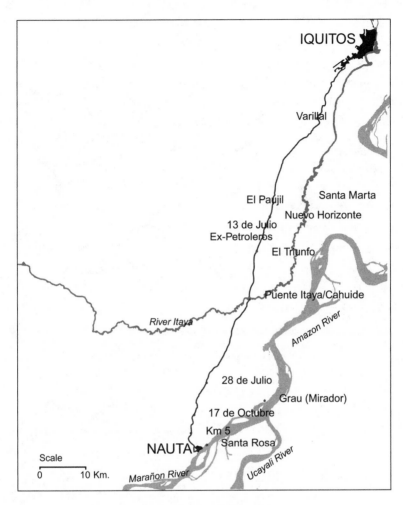

IQUITOS

Varillal

El Paujil

Santa Marta

Nuevo Horizonte

13 de Julio
Ex-Petroleros

El Triunfo

Puente Itaya/Cahuide

River Itaya

Amazon River

28 de Julio

Grau (Mirador)

17 de Octubre

Km 5

NAUTA

Santa Rosa

Scale

0   10 Km.

Marañon River

Ucayali River

Map 3. The Iquitos-Nauta road

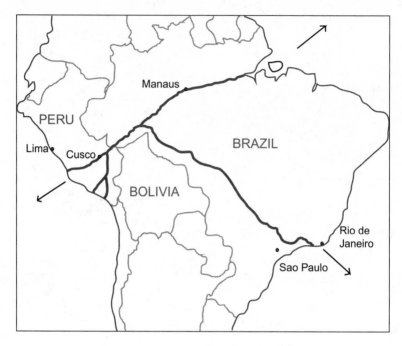

Map 4. The connective promise of Peru's Interoceanic Highway

Roads

INTRODUCTION

Anthropology, Infrastructure, and Expertise

Anthropology is a very different discipline today than it was in 1962 when Lévi-Strauss wrote: "The development of the study of infrastructures proper is a task which must be left to history—with the aid of demography, technology, historical geography and ethnography. It is not principally the ethnologist's concern, for ethnology is first of all psychology."[1] The modernist vision of a structural anthropology that could map a coherent web of cultural beliefs across groups has given way to forms of analysis aimed at documenting the fragmentation, rupture, and mutability of contemporary social life. And while anthropologists retain their commitment to learning from often marginalized dimensions of the social, they have refashioned ethnographic analysis in ways that take account of the instability of human relations in a complex, shifting, and often unequal and violent world. In short, the crux of the problem for anthropology has shifted. Rather than trying to understand the continuity of coherent cultural units, the focus is now on the social and cultural dynamics of processes of change. These processes are captured in a refreshed repertoire of concepts that range from emergence and encounter to confrontation, coproduction, and collaboration.

This turn toward analyzing processes of change, while pronounced, has left the discipline with an unresolved question regarding the specific contribution of an anthropological approach vis-à-vis other disciplinary perspectives. Although no longer comfortable with the stabilizing tendencies of the "conceptual structures" with which Lévi-Strauss was concerned, anthropologists continue to uphold the legacy of an attention to social relations, meaning, identity, and cultural differentiation. Indeed, this legacy can be seen as one of the crucial features that distinguish anthropological studies of contemporary processes such as globalization and neoliberalism from those conducted by geographers, historians, and sociologists. Moreover, this legacy has helped open the way to specifically anthropological studies of many new field sites—such as financial markets, biotechnologies, digital communities, laboratories, and bureaucracies[2]—sites we might characterize as exhibiting precisely the infrastructural qualities that were deemed to fall outside the remit of a former structural anthropology.[3]

This opening up of anthropology to the study of infrastructures, however, has had the effect of further unsettling the question of the project or purpose of anthropological description. This is true, at least in part, because the particular ontological configurations that infrastructures themselves manifest challenge some of the basic anthropological concepts that we have at our disposal for analyzing processes of social change.[4] Throughout the process of writing this book, we have endeavored to bear in mind both aspects of this methodological turn within the discipline of anthropology—that is, the *promise* that anthropological approaches hold for a distinctive and reinvigorated analysis of infrastructural relations and the *challenges* that infrastructures pose to the formation of anthropological knowledge.

When studying infrastructure the anthropologist must confront the problem of locating an ethnographic site without limiting the scale of description. Our decision to carry out an ethnographic study of roads in Peru responded to this methodological challenge. Since the 1980s Penny Harvey had been traveling to the small roadside town of Ocongate in the Peruvian Andes to conduct anthropological fieldwork on the politics of language. The road had always provided a significant backdrop to her studies of power and language, and it was clearly an important feature of the extended social geography of the area. People from Ocongate often talked about the opportunities that the road had brought, and the bilingualism of the population reflected extensive experiences of migration. Most men had traveled down the road at some time to earn fast money as contract laborers in the informal gold mining camps of the Amazon lowlands, or on the construction sites of the highland city of Cusco. Although the economy remained largely agricultural, many people also worked as traders, and a few even owned the trucks that took passengers, beer, and produce to Puerto Maldonado, or transported timber from the many illegal or semilegal logging enterprises back up to the highlands. But the road did not really come into view as a potential research site until Penny began to think more deeply about how to fruitfully conduct an ethnographic study of the Peruvian state, in ways that neither assumed prior knowledge of state form (e.g., as institutionally prescribed) nor discounted its distributed and abstract quality (Harvey 2005). At this point it became clear that the road could prove vitally useful to her in demonstrating how a mundane material structure registered histories and expectations of state presence and of state neglect.[5] Methodologically, however, this work remained focused on a single site, and attended to the ways in which the road materialized the state in Ocongate.

The possibilities for working at a larger scale arose during the early 2000s, when people began talking about a project that was to transform the rough, slow route from Cusco down to Puerto Maldonado into an "interoceanic highway." The new road was to link up with a transcontinental network that was already under construction, to provide a fully paved route stretching from the Atlantic coast of Brazil to the Pacific coast of Peru. Most of the residents of Ocongate were thrilled by the prospect that this highway would pass right through their town. They had long desired a new road. If these rumors were true, this major road construction project would finally deliver a fast connection to urban centers, cutting both the physical and social distance that undermined local hopes of economic prosperity. As the rumors were gradually substantiated, and it became clear that the project had attracted international funding, we began to explore the possibilities that this road construction project might hold for an anthropological study of contemporary politics. In particular, it was an opportunity to observe firsthand how an infrastructural project of this scale would, or would not, consolidate state presence in what was still a marginal and isolated region of the country (see map 1).

We thus began our research on roads in 2005 with the ambition to investigate how roads, approached as infrastructural technologies, might provide new perspectives on the politics of contemporary social relations. Ocongate was not the only village facing the prospect of a new road in its midst. Across the planet, roads are being built in developing nations as part of larger projects aimed at investing in infrastructural systems to support economic growth and modernization. Leading the way have been the so-called BRIC countries—Brazil, Russia, India, and China, which economists group together because they are all deemed to be at a comparable stage of economic development—but other developing nations are also involved, thanks largely to the fact that international lenders such as the World Bank are increasingly committed to offering development support via infrastructural investment. Capital investments in large-scale infrastructures are generally viewed as good business. Indeed, the short-term benefits of new labor opportunities and the longer-term promise of growth have led governments across the world (not just developing nations) to prioritize infrastructural projects in partnership with private capital as the primary response to global recession.[6]

Infrastructures have proven to be theoretically productive in the scholarly literature as well. Anthropological work on infrastructures began to draw together a set of diverse interests in technology, in material relations, and in material agency, fostering close relationships between science and technology studies and ethnographic approaches to the analysis of technological systems, social change, and development. The concerns of the growing field of science and technology studies have resonated with a general anthropological, philosophical, and historical interest in the human capacity to extend, reproduce, and energize relations—not least through their engagements with "things" and the ways in which such things are (or are not) made commensurate, equivalent, and exchangeable.[7] The focus on infrastructures as dynamic relational forms has begun to offer interesting analytical possibilities, allowing ethnographers to address the instabilities of the contemporary world, to highlight movement, contingency, process, and conflict in and through the study of particular infrastructural formations.[8]

It is also important to note that the social and political conditions that render infrastructures so visible in the contemporary world have led scholars to acknowledge the intrinsic heterogeneity of material forms—forms that previously seemed more solid, more secure, less precarious. This awareness of the vulnerability of infrastructures as public assets at a time of diminished state investment only serves to highlight the fact that there is continual material and political work to be done to keep these systems up and running.[9] Consequently, scholars increasingly view infrastructures as sociotechnical assemblages through which it is possible to tease out the arrangements of people and things and ideas and materials that make up larger technological systems. Whereas early work on technological systems emphasized the complex relational dynamics in play (Star and Ruhleder 1996) in any infrastructure, anthropologists have turned our attention to the political implications of the relational mix implied by infrastructural forms. Thus Anand (2011) calls infrastructures "the material articulations of imagination, ideology, and social life," while de Boeck (2012) refers to them as built forms around which publics thicken.

As we followed these currents in contemporary research on infrastructures, it became clear to us that our own work on roads would focus squarely on the form of the political. Indeed, our project originated from an interest in how to carry out a political anthropology that did not rush to assume the shape of the political, but rather made space for tracking the diverse ways in which politics is enacted, anticipated, and understood. We wanted to think about how the contemporary state emerges as a potent force in people's everyday lives, and at the same time not reduce the political to state presence, but also consider forces and agencies beyond those of state institutions and recognized political organizations. We thus began to pay more attention to the political purchase of infrastructural forms in their own right, the material politics through which they are brought into being, sustained, or undermined, and the ways in which they, in turn, are constitutive of political power.

Building on the foundational work of Star and Ruhleder (1996) we believe it is important to emphasize the "when" rather than the "what" of infrastructural formation. Their research points to the effort involved in creating infrastructural systems that "work," in the sense that they succeed in producing smooth flows, thereby obscuring all the complex relational mechanisms on which such flows rely. With this in mind, we explore the inevitable gap that exists between the intended effects of infrastructural projects and the way those intentions play out in actual practice. As we discovered, intentions always run up against the intrinsic multiplicity of infrastructural systems and the strength of the existing relational fields into which they are inserted.[10] Standards and regulatory frameworks carry a degree of compulsion in their ordering demands and routines, and yet they are always sufficiently open in their own internal differentiations to ensure that outcomes are unpredictable.[11] Lampland and Star (2009) emphasize the significance of this "invisible trouble," which is internal to infrastructural systems but remains hard to see and often not possible to talk about, appearing only in moments when the infrastructures appear to be at risk.

The expectation of "invisible trouble" orients us to the fragility of infrastructural systems and to those points of instability that can reveal crucial information about what and how infrastructures hold together.[12] This in turn reminds us that it is not simply the material presence of infrastructures (i.e., the paved surfaces of roads or the tracks of railways) that lead to the integration and differentiation of modern national territories and populations. Also of crucial significance are the expert practices, such as techniques of measurement, mapping, and description, that help shape processes of social transformation. It is in this spirit that we take seriously the planning and construction practices that go into the building of roads. Thus, for instance, our ethnographic approach to road construction focuses on the work that goes into holding things together, and considers the techniques and collaborations through which the world can take the form it does. In doing so we are extending Lampland and Star's (2009) characterization of infrastructures as fundamentally relational entities, and exploring how "invisible trouble" disrupts the plans and expectations of both the engineers and the wider population.

A focus on infrastructures as both virtual and actualized relational spaces thus allows us to open our deliberations on the political to include uncertain, ambiguous, and unstable outcomes. It also allows us to trace the habits, understandings, and entrenched assumptions that Lévi-Strauss had delegated to psychology, and reclaim them for a social and historical analysis of material relations. In development settings infrastructures are aspirational and carry great promise; yet they also carry threats of unwelcome change, of destabilization and increased vulnerability. They combine social memory and future imaginaries in complex ways that have to be worked out, as these temporal dimensions of infrastructural forms are also always heterogeneous. Our work on roads thus also attends to the imaginative and material practices that contribute to the emergence—the coming into being—of infrastructures as open-ended structural forms.[13] Included in this are multiple and often incompatible aspirations, technical procedures, organizational dynamics, employment relations, health and safety regulations, property regimes, and understandings of the "public good." It is through our detailed observation and description of these imaginative and material practices that we set out to explore how infrastructural forms provide such significant diagnostic spaces for contemporary political anthropology.

The Modernizing Ambition of Civil Engineering

In Peru, as in many developing countries, roads have meanings that go well beyond their physical functionality. Banks and governments view them favorably as technologies of social integration, economic development, and modernization. They are also relatively easy to sell to the wider public in places where poor communications systems exacerbate the physical and economic hardships of everyday life. Indeed, many Peruvians long for roads with a fervor that we found hard to fathom at first, given some of the more deleterious effects that roads are often held responsible for in more critical accounts. That said, it must be stated from the outset that people are not wrong in their understandings of the correlations between the presence of a road and the possibilities for economic growth that these smooth connective surfaces afford.[14] Roads enable the networked flow of goods, labor, and services. They deliver the basic conditions of modern living, although, as all scholars of modernity are aware, the benefits are uneven and unpredictable.

In the chapters that follow we trace the histories of specific road-building projects in Peru to explore the dreams and effects of past road-building programs as projects of national, and even of international, integration. As we learn more about the histories of these spaces, we should not be surprised to find that, in spite of their transformative ambitions, roads often fail to address the historically entrenched inequalities and injustices that render livelihoods so hard to come by and lives so precarious. Nor will it be altogether surprising to find that previous road-building projects have had all kinds of social consequences, which the new roads are somewhat paradoxically expected to mitigate. Nevertheless, roads continue to invoke, and to materialize, that fundamental principle of ordered freedom that underpins both modern politics and modern science.[15]

Although it is clear that roads have social consequences, the relationship between roads and their political effects is complex and often rather opaque. Roads as infrastructural forms manifest the political,[16] not just through the transformations that they promise but also by arranging and rearranging the mundane spaces of everyday life. Roads create spaces for institutional forms of governance exercised not just by elected state representatives but also by the inventions and interventions of designated experts (engineers, planners, architects, civil servants, doctors, scientists) and the devices and instruments they create and deploy (Mitchell 2002).[17] The practical and directed work of these technical experts is oriented toward the resolution of specific problems, which fold the social and the technical together to produce material rearrangements in the name of emancipatory transformation or "development."

The professionalism of civil engineers draws on a long history of ingenious intervention and pragmatic application of scientific ideas that rests on close observation of empirical phenomena and the ability to orchestrate the huge variety of components from which new structures are composed, and moreover to ensure that these components hold their form and perform their intended function. Much of what we have to say about the transformative promise of engineering in this book stems from our fascination with the tension in engineering practice between the stabilizing ambitions of the ordering impulse and the pragmatic acceptance that things are not always as they seem, that circumstances and conditions change, and that unpredictable things happen.

The ordering capacity of the civil engineer dominates popular and scholarly accounts of engineering work.[18] These accounts capture a key dimension of the modern engineer's own self-image, but as descriptions of engineering practice they run up against the limits of their own presuppositions and pay scant attention to the anxieties and internal critiques that have always been integral to modernist thinking. They also ignore the craft dimensions of engineering practice, a foundational craftiness that informs the particular ingenuity of the engineer. Science and technology studies and historians of engineering practice have tended to offer more complexity and show how engineering practice became so closely identified with modern statecraft. The work of these historians interested in engineering and the state[19] draws extensively on the work of Michel Foucault, particularly in their understandings of the ways in which normative configurations of power/knowledge stem from the modern knowledge practices enacted by state institutions. These knowledge practices have been shown to be key to establishing the importance of material practices to modern public institutions, which carried forward this particular combination of technics and moral improvement from the field of the "engine sciences."[20] Civil engineering expertise was thus central to the rise of the modern state and to its subsequent colonial elaboration in the nineteenth century, not only in relation to the technical infrastructures through which territorial integrity was imagined and materialized but also for the ways in which a sense of public benefit was consolidated around these very particular arenas of state practice.[21]

Contemporary critics of modernist development such as Scott (1998) and Mitchell (2002) identify the technical expert with such state effects. Indeed, when it comes to public works and infrastructural projects, the civil engineer emerges as the expert par excellence in these studies. Such expertise is deemed problematic precisely because it is exercised as a managerial practice and carries a hubristic moral force. External knowledge is brought in to situations from which something is deemed to be missing, and where correction or adjustment is deemed necessary to solve specific problems, again described and articulated in specific (partial) ways. From such scenarios it is not difficult to see why the engineer, as modern expert, emerges as the villain in the critical social sciences. The history of the modern state and the technology-driven developmental programs through which capitalist formations emerged and consolidated over the twentieth century have certainly achieved "progress" for the few at great social and environmental cost, producing what now appear to be irreversible toxic consequences that challenge our planetary future. As critics of modernist developmental programs have pointed out, the problem stems from the reductive effect of processes of standardization that are put in place to control uneven and unpredictable environments—controls that in practice are imposed to regulate and to enable the scaling up or generalization of productive or distributive processes through the rationalization of complexity. In such scenarios the specificity of local knowledge always appears inadequate to the task of producing a generic solution. For those more interested in how ordinary people deploy practical skills in their everyday lives, such generic solutions are by definition blind to human creativity, and hugely destructive of the relations through which such creativity is recognized and reproduced.[22]

Furthermore, the modes of authority through which the expert can assert an impartial and generic knowledge, variously referred to as "locationless logics" (Mitchell 2002), the "cultures of no-culture" (Traweek 1988), or simply the "God-trick" (Haraway 1991), have been shown by these critics to rest on very particular conditions of knowledge production and circulation. However, when one pays close attention to the practices through which expertise is performed, it becomes clear that the professional skill of the engineer always requires something beyond the abstractions of technical expertise. The mathematics and the metrics are necessary but never sufficient for the realization of engineering projects.

Those people whose work came to define the engineering profession at the birth of the modern nation-state were precisely those who were able to manage both social and technical complexity. Accounts of early engineering practice in the seventeenth century show, with the benefit of hindsight, how technical projects to transform the material world are replete with ideas and values about appropriate social relations, and thus show how it is that the material relations at the heart of engineering projects are intrinsically political.[23] Following from these historical studies of early engineering, our work explores the particular tensions that contemporary road engineers have to manage, and what these tensions can tell us about the contours of current social relations in Peru. Accounts that engineers give of their expertise appear on the face of it to simply focus on the working out of specific technical solutions, but such solutions necessarily always engage complex social values at the same time as they delimit the social world as external to engineering science. Moreover, the moral certainties and the ambivalences that engineering professionals in Peru expressed to us in the course of our ethnographic research serve as reminders that the engineers themselves are embroiled in the world from which they seek to stand apart in the exercise of their profession. This requirement of modernist development, to be both inside and outside, to have a reflexive awareness of one's own presence in that from which you stand apart, is a dimension of engineering work that the Foucauldian traditions pay less attention to, but which ethnographic study is particularly well placed to explore.[24]

In chapters 3, 4, and 5 we trace in some detail how this "double vision" of the modern engineer is played out in road construction projects in the never quite settled relation between regulation and its excessive externalities. Techniques for holding things stable emerge as central to this dynamic, returning us to the contemporary enactments of the engine sciences. As road engineers approach the world they seek to transform, they are required first and foremost to describe the world they encounter. In order to make a road a reality they must engage in the collection and ordering of samples, statistics, and measures, the drawing of maps, and the production and testing of models. All the devices of the engine sciences in their latest technological incarnation are deployed to help engineers fathom the essential characteristics of the spaces in which they find themselves. As the historians of engineering have taught us, these devices help the engineers to deal with complex realities; to know them in such a way that they are able to address the multiplicity, the heterogeneity of sociomaterial worlds; and to make an effective intervention, one that is visible, tangible, and enduring.

Such processes of description, however, are only one part of what engineers do. Once they produce a description that allows the configuration of a way forward, the move to implement the plan draws them back into the world of situated action (Suchman 1987), to the lived world of social relations. With this come compromises and the pragmatic accommodation of all kinds of people and substances that have an active role to play in the transformations the road will bring about. If historical studies have demonstrated the link between engineering and the emergence of the modern nation-state, then our aim in these chapters is to better understand the relationship between the contemporary configurations of expertise and the reorganization of contemporary political formations under conditions of neoliberalization. In paying attention to the formation and practice of engineering expertise as related to road building, we follow in the footsteps of previous ethnographers and science and technology studies scholars who have looked in detail at what experts do on a day-to-day basis.[25] Such approaches have alerted us to the heterogeneity of engineering practice (Law 2002b), and to the importance of looking at how the technical and the social aspects of engineering expertise are both differentiated and yet held together. This allows situations in which a notion of the engineering profession that involves the technical analysis of a preexisting natural world can coexist with the knowledge that all civil engineering projects are socially driven, and thus require a powerful and conscious engagement with the complexities of the social world.[26] The move from ordering to coordination involves a subtle shift in perspective with respect to the question of how roads emerge, and what they become. In the ordering frame roads appear as materializations of a particular mode of environmental control. Coordination, on the other hand, opens up our attention to the multiplicity of relations that must be negotiated to bring a road into being.

This doubling at the heart of engineering practice interests us, for it suggests a need to go beyond an analysis of the dominant politics of state engineering in order to attend to the intrinsic internal differentiations within engineering practice. Focusing on the falterings, the uncertainties, the unpredictable, and the contingent allows us to extend our understanding of road construction as a site of politics. Such a move changes how we approach the relationships between expert knowledge and state power. And while it in no way precludes a critical examination of how the material configurations of modern knowledge practices render the world knowable, and hence governable, it also directs us to keep a close eye on the effort required to effect contemporary modes of governance. Ethnographic attention to this realm of effort, and to its political ramifications, is where we seek to make our contribution. Regulation is devised and revised constantly in relation to an intrinsically unruly world, a world where imaginative possibilities inhere in the material, where the will to order is always undermined by the compulsion to imagine and to materialize the disordered counterpart. In short, we characterize our description of engineering practice not so much in terms of how technical knowledge is constitutive of state power through the enactment of mechanisms of control. Rather, we are more interested in asking how the work of coordination brings an affective dimension back into the mix, all the while acknowledging the uncertain and the unruly as integral to the politics of infrastructural relations.

Another aspect of our ethnographic focus on the work of coordination that civil engineers do is our attention to the ways in which the skills, orientations, and effects of contemporary engineering both acknowledge and erase other knowledges. We are keen to avoid setting up a simplistic opposition between local indigenous knowledges and generic modern knowledges, following Akhil Gupta's lead in assuming a certain hybridity to all knowledge forms (Gupta 1998). We therefore treat engineering expertise as equivalent to other knowledge forms, in the sense that we approach the engineers as people who deploy a range of knowledges, which in turn enables them to deal with multiple coordinates and adapt to a variety of different circumstances. Avoiding a simple opposition between different forms of knowledge is thus intended to highlight the relational dynamics through which roads emerge. It also allows us to consider the possibility that these same spaces result from a series of pragmatic accommodations, a drawing together of diverse materials, instruments, measures, forces, concepts, and imaginaries that might never quite coalesce to the degree that the ordering paradigm suggests, leaving a certain openness in relation to what such spaces might become and who might determine their futures.[27] Seeing road building as a coordinating practice provides us with another way of describing what we mean when we say we are exploring the material and social lives of roads.

Roads as Ethnographic Fields

As we stated at the start of the introduction, we conceived of our study not as a description of a particular place or space—that is, Peru—but as an ethnography of how infrastructures configure contemporary politics, taking full account of how the contingencies of the everyday are fully implicated in this process.[28] From the outset of our work in Peru we had expected that the roads would give us the opportunity to observe the negotiations between generic and specific knowledges, between those knowledges that explicitly emerge from place-based relations and those that carry the force of that capacity of expertise to apparently escape location. But ethnographic immersion led us to attend to the journeys that all descriptive accounts are launched on, the entities that they are mobilized to connect and bring into being, and those that they leave to one side and designate as irrelevant. Over time the road itself became the template for this mode of attention, and through the ethnographic process acquired its narrative force. Our approach to knowledge practices is thus not formulated as a critique of engineering expertise, and it is not intent on deconstructing particular claims to truth in order to reveal an otherwise concealed partiality. Rather, our focus is on how particular knowledges become credible and actionable, how they attain force and presence in the world, and how contemporary dynamics of power emerge in and through these knowledge spaces. Although many of those with whom we spent time on the road were keen to establish hierarchies of knowledge in ways that we will discuss in detail below, we did not set out to study self-identified communities (of villagers, occupational groups, protest movements, and so forth). We instead focused on the roads as spaces of projection and material transformation. In this respect the ways in which the engineers describe the land, its forces, its qualities, and its capacities, are treated as equivalent to the accounts we were given by peasant farmers, traders, or migrant workers. We approach the autobiographical narratives that explicitly drew connections between people and things as being parallel to descriptions that combined numbers, figures, and categories and circulated between the topographers and the designers as they sat at computer terminals in distant cities. We do not seek to establish a hierarchy of knowledge. We are more intent on engaging the specific orientation of particular accounts and in trying to work out what needs to be assembled to achieve descriptive adequacy in relation to such an orientation. The ethnographic task is to remain open to the possibility of multiple orientations and to track the effects of their coexistence.[29]

Early in our fieldwork much of our time was spent identifying the extensive cast of characters that would be part of our account. Road stories involve national, regional, and local government officials—from presidents of state to presidents of small communities, local residents, pressure groups, those identified as "stakeholders" and those whose presence was unmarked, or marked as illegitimate. The military and the police feature prominently in the histories of the roads we analyzed, as does the diverse group of experts working within private organizations and state institutions. The business of producing a road links transnational consortia to small local enterprises, and draws together professionals from many fields, such as engineers, designers, lawyers, economists, financiers, managers, journalists, archaeologists, and development workers from both secular and religious NGOs. The professional classes are themselves a small subset of a much larger workforce, equally varied in relation to skills and prior experience, which are played out through the differential opportunities that road construction affords men and women, old and young, the well-connected and those without influence. Moreover, swirling around the edges of the formal labor market are the layers of subcontracting, the travelers, migrants, squatters, and speculators, the hopeful, the dismayed, and the indifferent. If we now begin to gather the key nonhuman agents, our list extends exponentially through the categories of materials, instruments, machines, documents, and laws. The regulative landscape and affective desires that link these different groups and individuals are also components from which we seek to produce a description of roads as infrastructural technologies in order to learn more about how material practices and their resulting formations become constitutive of political power. Our task has thus been to adequately describe the way in which roads work as scaling devices, whereby we find state power in test tubes and measuring tapes, and global capital in the confrontation over the ownership of scrappy bits of land with contested histories.

Up to this point, we have focused primarily on theoretical matters that transcend any specific place. However, in no way has our intention been to diminish the importance of the particularities of Peru as an ethnographic location. On the contrary, we seek only to convince those who might have no particular interest in Peru that there is a great deal to be learned from what goes on there. As with all ethnographic projects, our ambition is to use the specificity of Peruvian history and the experience of road building in this particular place to draw attention to processes and politics that will be of relevance to infrastructural projects more generally. Our focus on the practices and transformations that inhere in road construction is taken as a means of embarking on a descriptive adventure that seeks to transform established ways of thinking about road construction and national development by thinking in new ways about civil engineering practice, as both an aspirational project and a material intervention.

Our story unfolds in three sections. The first section describes the entanglements of past desires and future imaginaries that emerge in relation to the process of road construction. Our central concern is to interrogate accounts of road construction for what they can teach us about how the relation between territory and economy takes shape and is transformed over time. Chapter 1 focuses on dreams of territorial connectivity as they relate to the ongoing project of state formation in Peru over the past one hundred years. Working with material traces of previous projects, oral accounts, and personal and state archives, we use these diverse and specific histories to build up a sense of how people locate themselves and others as participants in the movements of people, ideas, and materials that come to shape state space and public space in Peru. If chapter 1 focuses on the ways in which road construction projects have participated in the making of a particular imaginary of the Peruvian nation-state, chapter 2 focuses on how stories about people's experiences of road construction provide an alternative account of state absence and local entrepreneurial initiative. Focusing both on the way in which roads raise questions about land ownership, land use, and commercial life and the role that migration plays in the formation of social relations along these roads, this chapter provides an account of the ways in which roads participate in an emergent politics of differentiation.

The second section of the book takes up this tension between the promise that roads hold to create a collective future through a politics of connectivity and the intrinsic differentiating effects of these connective technologies, through an attention to the day-to-day practices out of which these political effects become manifest in road construction processes. In particular we attend to the establishment and use of the procedures and norms that guide engineering practice, and the articulation of these norms with a broader knowledge politics of road construction. Chapter 3 looks at the quantification of materials in construction laboratories; chapter 4 focuses on the health and safety procedures that govern the construction process; and chapter 5 takes the negative example of endemic corruption, or the apparent "failure" to regulate the roads as sites of wealth creation. Our ethnographic focus on the practices that are constitutive of each of these fields aims to interrogate the siting and shaping of the political, revealing the ways in which expertise is enacted through the mobilization of affective force and moral judgment, challenging the self-image of the expert as the disinterested and rational outsider. Our focus is thus not simply on the production of differentiated materials, populations, and transactions as sites of regulation. It is also on the work it takes to create these categories and to stabilize them as recognizable entities, work that we show to be dependent on the mobilization of highly charged affective relations.

The final section of the book considers the political ramifications of the modern project of social transformation via infrastructural engineering by identifying two new areas of attention for anthropological analysis of contemporary political relations. The first of these revolves around the figure of the "impossible public." Infrastructural projects are organized with an explicit awareness of both the positive and potentially negative impact of these projects on local populations. Much time and energy is invested in encouraging beneficial effects and mitigating negative effects. However, in the course of the negotiations between corporate/state projects and local communities, we find the generation of a mode of political action that appears to escape the conventional oppositions of political philosophy, for example, those between the state and the public, or between publics and counterpublics. We focus our attention on those people who respond to state projects in ways that fail to either accommodate or oppose them in ways that fit the expectations of the engineering professionals. We argue that in some ways these figures manifest the excesses of regulatory procedures as they follow alternative, uncooperative, or spontaneous possibilities to their logical conclusion. The unruly presence of citizens who refuse to behave responsibly manifests a troubling creativity and imaginative response to regulation. The politics of such responses is unclear. Such publics are not easily cast as social actors, yet their strategies are familiar, recognizable, and in many cases effective.

The impossible publics are mirrored in chapter 7 by our second key figure, that of the engineer-*bricoleur*, a further hybridized disruption of the narrative of modernist transformation. Rational, controlling, systematic, driven by political ideas of civil improvement and the power of man over nature, the engineer has provided an archetype against which other ways of acting and engaging with the material world are judged as more or less modern, more or less creative, more or less embodied and sensory. For many, this figure of the rationalizing engineer is mobilized to exemplify how it is that generic, universal knowledge is deployed—for better or worse—in a world that resists such coherence and unfolds in the complex dissonance and unplanned consequences of lived experience. Yet, as ethnographic subjects, the road engineers with whom we engaged in Peru complicate this image of reasoned action and universal value. Considering the methods by which specific engineering projects are demarcated from a more enduring process of infrastructural development, and in particular the means by which the completion of road construction projects is performed, we are returned to the practices by which engineers engage the uneven, unruly, and unstable environments out of which infrastructures are made. Engineering practice requires a minute attention to the relational world, which resists the disciplined divisions that characterize the dominant paradigms of scientific work. Things collapse into each other with interesting consequences. For while historically the professional engineer emerged as one able to manage the complexity of things, contemporary notions of expertise are more likely to focus on the shortcomings of engineering, specifically the supposed incapacity of engineers to recognize the social dynamics of technical relations. This paradox has inspired our engagement with road construction in Peru and offered us a way of thinking about how projects of infrastructural transformation enact and express the dynamics of contemporary political life.

PART I

Roads as State Space

Past Desires and Future Imaginaries

Chapter 1

Historical Futures

As we began our fieldwork in Peru in 2005, many people we talked to bemoaned the fact that Peru had never had a coherent roads policy. The road system was generally perceived to have been put together piecemeal and in response to short-term political goals. The roads that did exist were not well maintained, and those that were functional were not necessarily delivering prosperity to local people.[1] In some places the roads even seemed to enable illegal trade to prosper without building the modern economy they were supposed to foster.[2] The connectivity they afforded was channeling people and resources in aberrant ways. Conversations with people about the roads would rapidly turn to the problems that beset Peru: underdevelopment, social inequality, political violence, and the sense of abandonment experienced by so many. And yet there appeared to be no diminishing of the expectation that roads could and should deliver a better future. The problems were deemed to stem from the gaps and discontinuities in the infrastructural system, the uneven distribution of roads, and the failure to adequately connect major settlements or reach outlying areas. The solutions, commonly suggested, were a coherent development policy, rational planning, a long-term horizon, and a more sincere commitment on the part of the state to embrace its responsibility to the public good. Time and again we were told that roads were a basic requirement for producing an integrated, coherent, and democratic modern polity, and the lack of decent roads was a fundamental block to Peru's future prosperity. Indeed, we found that in general people were quite passionate about roads both as a prosaic answer to the abandonment and abjection provoked by inadequate provision and as a redemptive promise to bring about social change.

Peru is not a country without roads. Those the country is perhaps most famous for are its Inka roads, which are increasingly celebrated as valued cultural patrimony.[3] However, these exemplars of premodern engineering skill and political acumen are essentially footpaths built before the wheel arrived in Latin America. The modern road system grew with the advent of motorized transport in the twentieth century, which saw several ambitious attempts by successive governments to modernize and to integrate the diverse regions of Peru. The intrinsic environmental challenges of high mountains, dense rainforest, and the arid coastal desert have tested the skills of engineers who came to Peru from many parts of the world, often to try out new techniques or experiment with new materials. Military and civil engineering expertise grew hand in hand with the historical unfolding of political and economic relations that secured the support and the interest of those with the means to fund such ambitious construction projects. In this respect modern Peruvian road-building programs have always been complex affairs that required a mutual accommodation between the territorial interests of the state and the more fluid and less bounded horizons of international capital, trade, and expertise.

In this chapter we trace the history of these modern road-building initiatives in order to explore the ways in which roads have participated in the emergence of the globally interconnected territorial state. The chapter explores how the capacity of roads to conjure a sense of the potential of enhanced connectivity has both enabled the consolidation of singular sovereign national territories and participated in the production of multiple vibrant transactional networks. This then is a story of infrastructural development and the extraordinary promise that roads as connective technologies hold for both economic development and political integration. We tell our story from the perspective of two specific roads, each of which had been the subject of much hope and of considerable controversy. The first of these is Route 26, now commonly referred to as sections 2 and 3 of the Interoceanic Highway (see map 2). This seven hundred kilometer stretch of road, which runs from the highland town of Urcos to the border with Brazil at the Amazonian town of Iñapari, was the site of the major international construction project mentioned in the preface. The CONIRSA consortium, headed by the Brazilian engineering firm Odebrecht, had won the engineering contract and was undertaking the preliminary studies and design processes prior to the construction proper, which would not get into full swing until 2006. As a multimillion-dollar initiative, this road was the subject of considerable political debate. Supporters focused on the way in which the road would become a means of responding to the urgent need to attend to what had become an abandoned region of Peru. The promise of the project from this perspective lay in the possibility of a revitalized connection between these abandoned regions and Brazil, the dominant economy of Latin America. Critics, however, were dubious. Many in Lima were concerned that the project was likely to be of more benefit to the Brazilians than to the Peruvians. Rumors circulated that there was no proper plan, there were no adequate studies, and there was no rational attempt at integration. Furthermore, as the project progressed serious environmental and social problems, of which the majority of Peruvians living in the capital city of Lima were unaware, were increasingly aired on the national media. The region through which the road was to pass had long been the site of artisanal gold mining and illegal logging. However, it was only when the new, fast road focused people's attention on this region that they began to notice the scars in the forest, the mercury in the rivers, and the implications of an informal economy that sustained tens of thousands of workers and their families. The images that appeared on TV screens fueled a sense of connection to this distant place and showed it to be part of a shared national territory for which all citizens held some responsibility. Just what the nature of that responsibility was and how it was to be addressed goes to the core of our interest in infrastructures as sites for the analysis of political relations.

Our second road was no less controversial, despite the fact that it is only a one hundred kilometer stretch of highway (see map 3). Running between the city of Iquitos and the town of Nauta in the northern Peruvian Amazon, this road was apparently nearing completion in 2005 when we began our fieldwork. Colleagues working in the area warned us not to give too much credence to such predictions. The Iquitos-Nauta road had been halted by legal battles between contractors and the Regional Government of Loreto, and materials had been impounded. Although there were only five kilometers left to complete, some thought it could take years for matters to be resolved. With a seventy-year history of construction without completion, it was not surprising that many doubted that the new road would be finished soon. Over the years the road had contended with the clashing interests of loggers, hunters, environmentalists, and developers. To make matters worse, the road had an exceptional history of embezzlement and fraudulent practice. It is rhetorically claimed by some that when the actual costs are calculated per kilometer, this road emerges as one of the most expensive in the world. This is in spite of being situated in a marginal geographical location and reconnecting two places already linked by the waters of the Amazon River. The capacity of this short stretch of road to sustain this disproportionate degree of investment of time, energy, and money over the years in relation to its apparent lack of political and economic significance seemed to make it an ideal comparator to the Interoceanic Highway project, raising the question of how even the most unlikely infrastructure projects are able to sustain an ongoing emotional charge.

From the perspective of friends and colleagues in Lima, both of the roads, in their own ways, exemplified the problem with Peru's inherent lack of connectivity. The Interoceanic Highway was cited as a shameful example of political expediency. People commented that huge resources were being directed to an area of the country from which the majority of the population stood to gain very little and from which Lima, in fact, may well stand to lose.[4] The cost-benefit analysis did not appear to adequately demonstrate the need for this project. Feelings about the Iquitos-Nauta road were even stronger in this regard. The most expensive road on the planet was not even going anywhere in particular—it was a joke, even in Iquitos itself. As one person pithily put it, "The road was the dream of claustrophobics who didn't realize that they were living on the banks of one of the most powerful rivers in the world."

Yet, as we discuss in more detail in the following sections, the terms of such cost-benefit equations had never been straightforward. Throughout the history of these two roads, those in favor of them had brought social criteria into play that endowed these projects with the moral value of "public works" and the obligation of governments to attend to those on the margins by not allowing expansive territories of "abandonment." When we visited the towns and villages along the two routes we found long histories of enthusiasm for these roads. These residents were determined that they should not be deprived by those already enjoying the benefits of urban living in other parts of the country of infrastructures that were so necessary to local plans for economic and political development. Furthermore, although the history of road construction projects can be framed as the outcome of international negotiation and collaboration, in ways that we outline below, they were also, from a local perspective, the result of hard-fought political battles in which local people had engaged the state and campaigned for their rights to integration.

In what follows, our aim is to describe the histories of the two roads from the perspective of the contingent and overlapping perspectives of local campaigns and the longings they express for modernity and progress, as well as from the wider global histories of movements of people, skills, and technologies as they in turn have shaped the possibilities and prospects of transformation. We have been influenced by Timothy Mitchell's (2002) account of the emergence of the modern Egyptian economy, particularly by the ways in which he carefully traced the specific serendipitous effects of intercontinental flows and blockages, of wars fought elsewhere, of economic speculation, of expertise transferred, and of the undetected movements of insects, viruses, and plants. Mitchell excels at piecing together a larger story and showing how these global circulations come to shape the potential of specific places. We start at a different scale. We are interested in the "territory effects" (Brighenti 2010) of these two specific roads and in what they can tell us about how infrastructural relations simultaneously make national territories, international corridors, regional circuits, and specific localities. We are interested in the unexpected alliances and affiliations that emerge from different sources and at different scales, which together bring infrastructures into being and endow them with an aura of transformative potential. In this respect our interest in the histories of roads is not simply in what they have to tell us about the twists and turns of state formation, or even about the uneven experience of state presence. We approach the question of what kinds of spaces road construction projects produce in a more open way: we are interested in how these infrastructures come to "compose the spaces of which they are a part" (Allen 2011). Our purpose in this chapter is thus to show the multiple dimensions of these spaces with respect to how past dynamics of connection and disconnection configure and reconfigure relational futures.

The Interoceanic Highway and the Territorial Imagination

The route of the Interoceanic Highway project runs east from the southern Andean highlands toward the Amazon Basin where it continues for several hundred kilometers to the border with Brazil (see map 2). People had been traveling along this route long before it became named Route 26, long before a first hard surface was laid back in the 1920s, and long before it became the proposed Interoceanic Highway. From the early days of the colonial period frontier settlements in the Amazonian stretch of the route were focused on agriculture, rubber, and timber. Puerto Maldonado, named after the rubber trader Faustino Maldonado, who was active in the region in the 1860s, was a recognizable settlement by the 1890s and was officially recognized by the state in 1902. The town was a rubber boom port, focused on the Madre de Dios and Tambopata Rivers. It was thus a regional center, gathering and distributing products that arrived and left by river. However, since the rivers in this part of Peru all flow into the Amazon Basin, there was no territorial continuity with the Andean hinterland. It was quicker to get to Lima via Bolivia, or along the northern tributaries of the Amazon and from there to the coast, than it was to attempt the journey by going directly from east to west.

To remedy this situation, the Peruvian state offered vast concessions to private companies in exchange for the construction of basic infrastructures such as roads and bridges. The U.S.-owned Inca Mining Company was given two million acres of land along the Tambopata River in return for a commitment to build a mule road from Tirapata (a town on the high-altitude railway out of Cusco) to Puerto Markham (the highest navigable point on the Tambopata River).[5] However, despite the enthusiasm for roads to facilitate the flow of rubber from the remote regions of Madre de Dios, there was no inclination for any form of political integration at this time. Indeed, in 1912 Madre de Dios was explicitly differentiated from the Cusco region and granted departmental status in its own right.

Over a century after Peru declared its independence from Spain (in 1821) transnational territorial imaginaries were far from stable, and in the first decades of the republic territorial politics was directed to securing the frontiers. The demarcation and defense of national borders involved both diplomacy and skirmishes with the newly forming nation-states of Ecuador, Colombia, Brazil, Bolivia, and Chile in areas where there was little possibility of an established state presence. Securing national territories at the borders involved strategic decisions about how and where to attempt to stem the flows of goods, persons, and capital in ways that allowed the newly formed state to exercise its sovereign power without having yet produced the infrastructures through which it might produce effective territorial orderings.

The story of the rubber boom exemplifies some of the key problems facing the Peruvian state at the turn of the century and helps provide background to some of the early forays into road construction in the regions where we worked. The rubber boom was well established in the northern Amazon by the mid-nineteenth century. Peruvian rubber was highly valued in international markets. It was used to manufacture the world's first pneumatic tires, produced in France by Michelin in 1885, and provided the raw materials for the new industrial products developed by U.S. companies such as Goodyear, Goodrich, and Dunlop.[6] However, with no roads connecting the coastal cities to the interior, there was no easy way for the Peruvian state to secure revenues from the commercialization of this product. Accounts of the rubber boom are accounts of unregulated extraction, extreme violence, slavery, and immense profits for the few who managed to control the trade.[7] The city of Iquitos grew from the spoils of this business, and so too did the fortunes of those living in the area of the southern Peruvian Amazon through which the Interoceanic Highway now passes. In the late nineteenth century the rubber barons Carlos Fitzcarrald,[8] operating out of Iquitos, and Máximo Rodriguez, a Spaniard who controlled much of the area around what is now the border town of Iñapari, aggressively worked to secure control of these lucrative territories alongside, and against, the Bolivians Nicolas and Francisco Suarez and their rival Vaca Diaz on the southeastern frontier of Peru. Rodriguez built the first road in the region. In the late nineteenth century the Peruvian state was in no position to secure the control and regulation of its Amazonian territories as it fought a losing battle with its Chilean neighbors in the War of the Pacific, ceding extensive coastal territories and along with them the rich guano deposits. At the turn of the century there was no possibility of public investment in the kinds of infrastructural projects required to sustain extensive exploitation of frontier resources. Nevertheless, some areas were more accessible to state intervention than others. Thus, in the Department of Loreto in the northern Peruvian Amazon (where the Iquitos-Nauta road now runs) the state did have a presence that enabled the enforcement of customs and excise regimes in ways that secured a channeling of the rubber boom prosperity into the regional and national economy. However, in the southern Department of Madre de Dios there was no such state presence.[9] Although neither region had roads at this time, the crucial difference between them had everything to do with connectivity. In the northern Peruvian Amazon, the river is navigable and flows through Brazil to the Atlantic coast. In the region of Puerto Maldonado the rivers offered no such possibility, and despite the efforts of rubber barons such as Fitzcarrald to find ways to connect across river systems, no such connectivity was achieved at that time.

In the Andean region there had been initiatives since the 1860s to invest in the construction of railways that would enhance connections to the coast. Before assuming the Presidency in 1872, Manuel Pardo had published in the intellectual journal *Revista de Lima* on the need to invest income from the diminishing reserves of guano into national infrastructures that would support both mining and agricultural production. However, by the second decade of the twentieth century road construction had begun in earnest. The first major road to the interior of the country, the Carretera Central, was initiated in 1918 and soon stretched 231 kilometers from Lima to Tarma. President Augusto Leguía, in office from 1908 to 1912 and from 1919 to 1930, was a modernizer, committed to enhancing national integration through the systematic implementation of a road construction program. He was not averse to deploying authoritarian means to bring about the changes he was looking for. In 1920 he famously passed a law of conscription that obliged men between the ages of eighteen and sixty to work on the construction and repair of roads in the province in which they lived for six to twelve days a year.[10] The law provoked considerable social unrest. However, recent research suggests that the protests and confrontations were not directed at the ambitious road-building program itself, or even at the law that required people to work for free on the roads. The problem was that across the country the landowning classes, assuming their ability to act beyond the reach of the state regulation, used the law (which they administered) to conscript labor for other purposes or to force people into working at times that were detrimental to their own agricultural or trading concerns. The road-building program itself was always popular: it was one aspect of a desire for connectivity that expressed a yearning for a stronger central state presence to counteract the abuses of local elites.

The 1920s was thus an important period of social change in Peru in which road building played a key part. The hold of the aristocracy was diminishing, and the postwar era in Europe and the new technological dominance of the United States was opening up many areas of social and cultural life. In Leguía's second period in office (1919–30) eleven thousand kilometers of road were built in an explicit attempt to penetrate the enclaves of local elites through the implementation of public works.[11] The enthusiasm for road construction in the 1920s was thus primarily a story of the engagement of local people who supported roads as a technology of an emergent modern state that might liberate them from the colonial servitude of established elites, an ambition in some ways reciprocated by a state eager to exert its presence. Through these policies roads came to be firmly associated with an ambition to integrate, modernize, and civilize.

The ambition to civilize through integration was not just a domestic Peruvian project, for it resonated in important ways with broader U.S. designs on South American transport systems at the time. U.S. foreign policy at the beginning of the twentieth century had set its sights firmly on the possibilities inherent in unifying transport systems that might realize dreams of hemispheric integration and foster the dissemination of U.S. cultural values, goods, expertise, and political aspirations to all of South America. The construction of the Panama Canal (1904–14) is probably the best-known example of spectacular engineering in this vein, but roads were also key to dreams of transcontinental transport infrastructures that would extend U.S. influence throughout Latin America. In 1919 the Pan American Commercial Conference was dedicated to the promotion of U.S. building methods to the Latin American states. According to Ricardo Salvatore:

> Penetrating Latin America through roads was functional to an imperial vision that presented technology as an instrument for hemispheric integration. Selling cars and building roads were means to exhibit to Latin Americans the achievements of a technologically superior society. Hence the project of an intercontinental highway encapsulated both foreign-policy dreams and business objectives. It projected into a mechanical package (automobiles + highways) the expectation of the American middle class and the political dreams of its leadership. At last, "Latin" and "Anglo-Saxon" America could be united by a technology that was representative of American mass consumer culture. (Salvatore 2006, 677)

This utopia was never realized, but that is not to say that it failed to leave its mark. From the 1920s onward the area that the Interoceanic Highway would later pass through began to change, and one of the most significant changes was the gradual move away from the rivers and onto the roads. Puerto Maldonado led the way. Two devastating floods in quick succession in the 1920s saw a planned and orderly move away from the river onto higher ground. Here a modern city, with squares, avenues, and public buildings, was built by the residents. By the 1930s mule trains were regularly traveling between Puerto Maldonado and Cusco, and travel, while slow, was continuous. The railway that runs down the spine of the Andes between Juliaca and Cusco was completed in the 1940s, and there were plans for a line to connect to the navigable reaches of the Madre de Dios from either Tirapata or from Urcos,[12] along a route that would, in time, become Route 26, the Interoceanic Highway. In the 1930s there was still much expectation that railways would form the backbone of the Peruvian transport and communications systems. The first bioceanic line had been completed in 1910, linking Buenos Aires on the Atlantic coast of Argentina to the Chilean capital of Santiago on the Pacific coast. President José Pardo, following in his father's footsteps, imagined a rail network that would interconnect Peruvian territories. However he had aspirations to intercontinental connectivity. His plans were for a major railway along the length of the country's western coastal fringe (where the Pan-American Highway now runs), another along the eastern side of the Andean mountain range, and five interconnecting lines across the Andes. The grid would then be filled with a network of roads to extend the infrastructural reach.

During our time on the Interoceanic Highway, we were introduced to an elderly man named Benedicto Kalinowksi who had been involved in the attempted construction of a railroad that would connect Puerto Maldonado with the highlands. Don Benedicto was ninety-three years old. He was born before the road arrived, in the settlement of Cadena, an isolated outpost established by his explorer father between the towns of Marcapata and Quince Mil, and his name still resonated in the many stories we heard up and down the road. He told us from firsthand experience what life had been like in this region before the road arrived.

Don Benedicto conjured an image of himself as a child of adventure, living through what the jungle produced for him and for the encounters it generated. His father, Juan, had come to Peru in 1880, sent by a Polish count to collect samples and to work as a taxonomist. His passion for the exploration and classification of South American plant and animal species was later taken up by another of his sons, Celestino, who is now recognized as one of the founders of Peru's Manu National Park. Don Benedicto was born in the house that his father constructed after having traveled with his wife, a mule, and some dogs into what was then a remote area of impenetrable forest. For thirty years his father was the only landowner in the vicinity, living a solitary life with just his wife and his children. He stayed in this remote outpost purely for his love of the biodiversity it offered him. Meanwhile, Benedicto and Celestino grew up alongside their siblings in this remote space, which was over a week's travel by foot or by mule from the nearest urban center of Urcos.

As he started to talk to us about his life, Benedicto began to weave a beguiling tale of a life lived as a diplomat of that frontier, a conqueror of the wild, and a seeker of hidden treasures who gravitated toward and traveled with missionaries, explorers, tradesmen, and engineers who found themselves working in a place they struggled to understand. He had always also been fascinated by the hidden riches of the forest, and he recalled how he and his brother went trekking into inhospitable terrain in search of pre-Hispanic ruins and Inca gold. He explained about the logging industries and how he had tried every way possible to make a living from gathering mahogany, cutting it in his sawmill, and transporting it to Lima, but the business suffered from the absence of roads. Everything had to be moved either slowly by mules and on foot or by incurring the expense of using planes and helicopters. He also reminded us about the contraband and the corruption and of the frustrations he experienced with trying to work with people who came from the United States, from Lima, or even just from Cusco to transform the jungle with no understanding of what living in these places involved. He then told us how he had worked for nine months for SINAMOS [13] the state agency charged with implementing and supporting the Agrarian Reform of 1969.[14] His particular duty had been to help plot the route of the proposed Pacific Railroad, a project that Don Benedicto remembered as a rationalized route cutting a straight line from the Brazilian border over mountains and across rivers to the city of Cusco.

However, Benedicto recounted how the work that he did was never realized, recalling how the plans he had helped to draw up ended up being destroyed in a fire at the SINAMOS offices in Puerto Maldonado. The proposed Pacific Railroad was not the only railway that failed to materialize.

The history of rail construction in Peru was beset by financial and political problems, and by 1930 the British-owned Peruvian Corporation controlled the rail interests. In addition, the Peruvian state was severely in debt to the Corporation. Furthermore, the railways had been built—with foreign capital and expertise—with the clear aim of facilitating foreign exports, rather than with any intention of fostering domestic markets.[15] Thus it was that, in the 1930s, as ever-increasing numbers of motor vehicles were arriving in Peru, rail ceased to compete for the limited public funds available, and the option to develop road infrastructures was deemed both politically and economically more viable.

Work had begun on the penetration road out of Urcos in 1922. Moving out from Cusco, the road reached the town of Ocongate in 1936, Marcapata in 1940, and Quince Mil in 1942. As in the lowlands, the Andean highlands were controlled by the owners of the huge estates. Ocongate lay just outside the huge estate of Lauramarca, and residents were not beholden to the demands of the *hacendado*, although the *hacienda* still controlled the market. The arrival of the road was hugely significant in the history of Ocongate as it gave people the basis from which to assert their independence. Although initially used to the advantage of the *hacendado*, who called up troops from Cusco to put down the challenge to his control of the region, the road is remembered today as a modernizing force. After the road arrived, more traders appeared, school teachers arrived, and the town began to build its own self-image as a modern political center that stood in explicit opposition to the authoritarian control of the landowner. The townspeople did eventually succeed in taking control of the local market, in large part because the road afforded them a privileged alternative trading base. In Marcapata, the site of the next great hacienda on this route, the landowner ensured that the new road bypassed the town and served the trading interests of the estate. It was only many years later, after a fatal landslide killed road workers and blocked communications with the lowland valleys, that the townspeople managed to draw the road into the heart of their town, an advantage that they energetically defended against all the intentions of the contemporary engineers to contemplate alternative routes.

By the mid-twentieth century powerful landlords, including some foreign companies (notably, the Inca Rubber Company), controlled vast tracts of land in the Andean lowland (*ceja de selva*) area of what was to be Route 26, where they also cultivated brazil nuts and traded in timber. The landowners themselves were isolated during that time, and one of our interviewees suggested that they might well have built roads just to be in contact with each other, to see and be seen by each other (*para verse la cara*). There was little or no state presence, and the workforce of the estates, which was often unpaid and cruelly treated, was at the mercy of the landowners. We were told of how native people were rounded up and forced to work, and how they were frequently enticed into debt relations from which there was no escape. Others were simply sequestered and not released, including a community of workers from as far away as Ecuador, who were discovered in the region in the 1970s, much to the shock of the administrators of the Agrarian Reform.

Slowly the road followed the people, in turn accelerating the migration process. By 1941 it had reached the far bank of the river in Quince Mil. Quince Mil was a new town that grew up alongside the road. It was here that we met Don Braulio Reina, who remembered coming to Quince Mil in the 1930s. Don Braulio was born in Marcapata in 1927. His father had come south from northern Peru, looking for work, and his mother was from Urubamba in the Cusco region. Together they had set out from the highlands of Cusco looking for a better life in the haciendas on the eastern flanks of the Andes. Don Braulio knew the names of all the haciendas that had been operating at the time in the area; he heard people talking about them and had visited them as a child. His family had moved from Marcapata in response to a government colonization program, and they had come down to Quince Mil with a group of Russians. The Peruvians and the Russian colonies had been set up on opposite banks of the river. It is not clear from Braulio's account when exactly this happened, but he said he was crawling by then. There was nothing in Quince Mil when they arrived. It was virgin forest, which the colonists were expected to bring under cultivation. These early settlers, who were largely anonymous in Don Braulio's account, were followed by people with family names such as those we encountered on our first journey down the road: Zlatter, Meza, Stambor, Kalinowski. These were men of a different social class, men who came to make money from the area in a more ambitious way, men who were remembered as individuals, men who left material traces in the landscape—old machinery, abandoned houses, plots of land that their descendants still lay claim to. Some of them were Peruvians, but there were also Yugoslavs, Russians, Poles, Japanese, and Americans. Stambor and Zlatter set up sugarcane plantations and manufactured the sugarcane liqueur, *aguardiente*. They drew more people into the region to work, with some coming hundreds of miles on foot. By this time, the place had shops and a name. Don Braulio was about ten years old, and the town was called Kimsa Challwa in Quechua, or "Three Fishes." The name was an ironic dig at the landowners. On Sundays Zlatter and Stambor would come down to the river from their haciendas, looking for gold and keen to hunt and fish. On one occasion they used dynamite to fish, but after the explosion only three small fish floated to the surface. The name Kimsa Challwa registered their petty failure.

By now this was now a major trading route, with mule trains of up to two hundred animals bringing goods down from the Andes to service the miners who were flooding into the region. There was work loading and unloading the animals. Goods for the shops were winched across the river on pulley systems while the animals swam across below. However many things were brought in, there was never enough because there were so many people now living in the area. These were times without authority, law, or security. And they must have been times of deep desperation for many as well. To finish his story, Don Braulio told us how the Quechua name Kimsa Challwa (Three Fishes) was replaced by the Spanish name Quince Mil (Fifteen Thousand), which sounds quite similar although it registers a different number and a different story of failed enterprise. This time it referred to a Russian who had come looking for gold with machinery and all sorts of equipment. However, he was looking in the wrong place, and he lost his fifteen thousand soles, which was everything he had.[16] When we asked about the descendants of these Russian colonists we were told they had moved on as fortunes and markets changed.

As we continued our journey we found similar stories of an initial isolated settlement, the eventual arrival of the road, and, following the road, a new influx of people, energies, and dreams. The roads followed the people, and people followed the roads. After Quince Mil, Mazuko is the next major settlement. This town is the center of artisanal gold extraction today, but it takes its name from a market gardener, Jorge Mazuko, who came into the region in the early 1940s. He traded vegetables to the dispersed Andean settlers in the region, who were arriving in large numbers in search of gold. He did not work the gold himself, but he acquired it by bartering his vegetables, paddling up and down the rivers in a canoe. As in Quince Mil, the naming of the town does not relate to a moment of state foundation or recognition. Rather, it simply registers, in a somewhat arbitrary way, the presence of one of the people who was there before the place had become identifiable as a specific location or destination to which one might arrive. Mazuko's base on the banks of the river gradually became established as a point of arrival and departure, a place that, as it became more populated, became somewhere people that could come to, to get supplies and to sell or barter their gold.

When the road arrived in Mazuko it shifted the axis of activity away from the river. The road brought a different kind of opportunity and created a new kind of settlement. The first settlers of the current town of Mazuko— which is situated several kilometers from the port on the riverbank—were people who were working on the road and had not returned to the Andes. They joined the artisanal gold workers, and in general were people of modest means. But with the road came the first major capitalist intervention in the region with the arrival of Barracudo, named by many as the man who had really established their town. Interestingly, in terms of our current narrative, Mazuko himself was primarily a man of the river, constantly on the move, distributing goods, and bartering and trading with people on local terms. Barracudo's intervention was of a different kind. He was a man of property, he had six hundred head of cattle, he was of the generation of investors who saw possibilities in the transformation of forest into pastures, and he had established himself at the center of a trading empire. As you walk around the center of Mazuko today, people can point out to you the extent of Barracudo's property. His inhabitation of the forest was clearly modeled on the landed estates of his Andean counterparts in his home region of Apurimac.

The road did not reach Puerto Maldonado until the 1960s. Here, unlike the itinerant trading settlements of Mazuko and Quince Mil or the hacienda towns of the Andean region, the road arrived at a thriving river port. Puerto Maldonado had prospered and had developed its own modes of connectivity in the years of the rubber boom. The all-powerful landlords, who had once brokered their independence from state control in exchange for infrastructural provision, had to accommodate to new, more distributed interests in the decades following the Agrarian Reform of 1969. Major landholdings throughout the country were restructured as cooperative enterprises and managed by technical experts who were intent on modernizing the economy. The energy of this period was focused on the modernization of agriculture rather than on infrastructural improvements per se, and the next great "road-building" program did not get under way until the 1980s, under the enthusiastic guidance of President Fernando Belaúnde Terry who, in his second period in office (1980–85), embarked on a major program of public works that were explicitly designed to unify the national territory (*unir la patria*). President Belaúnde Terry was already famous for fiercely promoting the construction of a communications route along the eastern side of the Andes during his first period of office in the 1960s. His dream had been to open up the east of the country and link the productive potential of the Amazon region to the populated coastal zones. The road, known as the Carretera Marginal de la Selva, was a huge engineering and diplomatic endeavor. Although very little of this road was actually built, the modernization of Route 26 was part of his wider ambition to build this great Amazonian Highway, which he imagined would provide a new connective possibility by which the producers of the Amazonian region would provision Peru's coastal region with meat and oil.

Belaúnde famously traveled around Peru on a donkey and knew what it meant to travel in places where roads were narrow and unsafe, and where the lack of bridges made the transport of goods a lengthy and labor-intensive process. His speeches celebrated the work of engineers and road construction workers as the anonymous heroes of Peru. He declared that the new bridges would bid farewell to the rafts routinely used to carry heavy vehicles across rivers, although at Puerto Maldonado rafts were still in use in 2010 as the Interoceanic Highway waited for the completion of the last major bridge, the 722.95-meter Billinghurst suspension bridge. The bridge had been commissioned and delivered in 1985, but it had sat for over twenty-five years in a hanger at the Puerto Maldonado airport, awaiting the moment when political will and financial possibility might coincide to enable the completion of this final link in the road from Cusco to Brazil.

As the bridges appeared in the 1980s, airfields that had existed at Quince Mil, Mazuko, and Iberia closed. Transport routes were firmly established, although the road itself was not maintained. By the mid-1980s, when Penny first started fieldwork in Peru, it was possible to drive a truck or an old bus from Cusco to Puerto Maldonado. The journey would take anywhere from two to ten days depending on the weather and the state of the road. Landslides and broken vehicles often shut the road for even longer. The lack of passing points had led to a timetable of alternating access, one day in one direction, the next in the other, but breakdowns and delays complicated things, and trucks often had to pass each other on narrow and unstable tracks. Many people died on these roads, and the journeys were made only when absolutely

Figure 1. Barges crossing the Madre de Dios River

necessary to transport goods or to get to a place of work. In this respect the road, although more or less continuous (with some raft crossings), provided only a weak sense of connectivity along the route, and the precarious conditions led people to think more readily in terms of abandonment and longing than in terms of integration and inclusion.

President Alberto Fujimori brokered the funding for the completion of the road after a memorable visit to Ocongate in the 1990s, when local people had lobbied him directly to bring them a safe, all-weather surface. The timing was good. Fujimori was under pressure from the Brazilian state, which was waiting for the Peruvians to upgrade the road. The Brazilians had built a fine new highway up to the border with Peru on the understanding that they would be able to connect to the new superports on Peru's Pacific coast as soon as the Peruvians were ready to fund their part of this transcontinental route. The Peruvian government was thus under pressure to sign the agreement to fund the project to build the Interoceanic Highway, and it finally did so in 2004. Despite opposition to the capital expenditure that this road represents (estimated as over US$1.5 billion), the government approached the project as an infrastructural investment with the potential to foster both economic growth and a more general sense of welfare and inclusion for this region of Peru. The argument was that Brazil would use the road to export soybeans and cereals to Asia, and Peru would charge Brazil for the use of the Pacific ports. At the same time, the modern, all-weather surface would facilitate communications through one of the more underdeveloped regions of the country, and in turn it would open new possibilities for the export of Peruvian goods such as cement to the southern states of Brazil.[17] Skeptics raised concerns, not the least relating to the twenty-five-year franchise offered to a Brazilian construction company to operate tolls along the new route. This same company had interests in the design and construction of new ports and hydroelectric projects, which this new road would also link up, creating what could become a space of undoubted connectivity, but one that was effectively annexed to Brazilian commercial interests. The diplomatic efforts made to counter these criticisms have rehearsed what we have seen is a deep-seated understanding of the social value of roads and of connectivity in general. Elena Llosa quotes the Brazilian ambassador to Peru, Raul Fernando Leite-Ribeiro, as saying, "We are not simply attempting to construct a road between Brazil and Peru. The idea is to create a broad and complete program of integration and development in the whole region. The road would function as a kind of spinal column. . . . Firstly there is a concern to promote the development of the south of Acre and Madre de Dios; secondly to integrate regions; thirdly this project will give both countries a connection to both oceans [*bi-oceanidad*]" (Llosa 2003, 19–20).

By the time construction began in 2006 a basic route had been defined, which for reasons of cost was not to deviate in any substantial way from the existing road, a road that as we have seen was largely built to connect the landholdings of the powerful in the first half of the twentieth century. However, while this frustrated some observers who dreamed of a new "rational" route through the Andes, it was also the case that although the landlords no longer held sway, the larger towns in the region were by now settled along this road. Local residents feared and fiercely campaigned against rerouting. Despite many prolonged tests and negotiations, in all cases they succeeded in ensuring that the new road did not bypass existing roadside settlements of any size. Ccatqa, Ocongate, Marcapata, Quince Mil, Mazuko, and Puerto Maldonado remained the key roadside towns.

This brief overview of the history of the Interoceanic Highway reveals the social entailments of this infrastructural project and some of the complex "territory effects" of an emergent Peruvian state. Despite a history of state entanglement, this road does not manifest state presence in any straightforward way. On the contrary, for many decades it exemplified state absence, revealing how at times the state had little capacity to exert any control and was obliged to negotiate with the economically more powerful frontier entrepreneurs, who effectively controlled those regions of the country that were less accessible to the authorities from urban centers. At the same time, as Llosa's quote from the Brazilian ambassador illustrates, the road has continued to evoke a powerful image of state integration. We will encounter this again in northern Peru: the idea that roads are the central physical support and nerve center of a regional economy that would create progress and development through closer integration and establish links to the Atlantic and Pacific oceans, enabling trade to assume global reach. The trope of interoceanic connection has been particularly important here. When considered in terms of a national road-building project, the Interoceanic Highway is of limited consequence, extending into what is for many an empty space, connecting one of the poorest regions of Peru with one of the poorest in Brazil. However, when conceived of in terms of Latin America as a whole, and specifically with respect to Peru's relationship with Brazil, this tentacle into the jungle represents a potential link to networks beyond the immediate territories through which the new road will pass. As the ambassador was keen to emphasize, the road marks for Peru the possibility of a link to the Atlantic coast of Brazil and to markets in Europe and beyond, and for Brazil it represents an opening to the Pacific Ocean and to the growing markets of China and the Far East (see map 4).

The road thus emerged in ways that conjured the presence of the state (or states) and its absence (in the experience of utter abandonment by those who wished or hoped for some kind of active administrative or legal force). Even with this one road it is clear that the general enthusiasm for road construction at the beginning of the twenty-first century was motivated by many different concerns and addressed the very uneven and place-specific modernities that the prior roads of the twentieth century had already produced. Far from creating a homogeneous and integrated territory, these early road construction projects had entrenched a sense of discontinuous space and differential capacities for moving around. The strong contemporary desire for a new road was lodged in this prior sense of abandonment and marginality, which was explicitly conjured in the development programs that continued to promote road construction as a means to economic growth, increased welfare, and a strengthening of national integrity. This story was echoed in our case from northern Peru, where we also find the history of the road connected to the ambiguous presence of the state. However, the story of the Iquitos-Nauta road above all revolves around a sense of thwarted modernity and a passionate desire for connectivity that has been repeatedly undermined by the other possibilities that infrastructure projects put into play.

The Iquitos-Nauta Road: Rhythms of Hope and Disappointment

Nauta is a small town (population just under twenty thousand) situated on the banks of the Marañon River, just a couple of kilometers upriver from where the Marañon and the Ucayali Rivers meet to form the Amazon. People from Iquitos talked somewhat disparagingly of the place. "Where do you get to if you go down the Nauta road?," we were asked, and promptly informed, "to the end of the Earth." Not surprisingly Nauta residents took a different view. Their narratives emphasized how in previous times Nauta was very much at the center of things. They told us that Nauta was the first native urban settlement in Amazonia, founded in 1830 by Manuel Pacayá, chief of the Cocama people. They showed us the earthenware pots that stand in the central square of the town and explained that in Cocama the word for "pot" is *amauta*, an etymological connection that invokes an important strand of Nauta's sense of historical origins, and its claim to indigenous authenticity. However, the place name also invokes the nautical history of the town and the important association with navigation and shipping that seems to underpin contemporary desires for the road connection. The financial fortunes of Nauta traders had flourished in the mid-nineteenth century when Nauta became the most important port in the Peruvian Amazon, a key node in a regional economy that was radically transformed by the rubber boom. It was the final duty-free port of the Amazon River system, and it effectively marked the entry point to Peruvian national territory for ships and goods arriving from Europe and Brazil.

Over the following decades the people of Nauta began to imagine their town as being on the brink of modern prosperity. Connected to global markets via the cities of Brazil, Nauta was set to flourish at just that time when, in this region at least, state boundaries had begun to stabilize and attention had turned to the possibilities of integrating a national territory. We were told how the people of Nauta were ready to play their part in building the new Peruvian nation. However, Nauta's potential as an important center of Peru's new modern economy was never realized. In 1930, just as this possibility began to materialize for the families who had built their businesses there, the river moved. Nauta was left without a viable port, and dreams of modernity and prosperity collapsed. The first attempts at road building in this area were thus not the roads of national integration but desperate efforts to try to maintain some viable connection to the river. But engineering such structures across shifting sands was beyond the reach of the townspeople, who were forced to watch the city of Iquitos grow exponentially while their small town remained a satellite settlement, some twelve hours away upriver. Charged with the energy of the unjustly defrauded, residents of Nauta began to campaign for recognition and some restoration of their capacity to continue their involvement in international trade. The aspiration, born when both Iquitos and Nauta experienced greater connectivity to Europe than they did to Lima, never faded.

Iquitos had a longer history of wealth, and today it is known for the elegant boulevards, squares, and the faded glamour of the exquisite mansions of this thriving city. These physical monuments to past riches exist alongside the stories of boats arriving with tall, beautiful, and exotic people, and of incredible parties on which the rubber barons lavished their fortunes. The parallel tales of slavery, murder, and suffering detach from the city—and linger in the forest and in the past. Iquitos is now Peru's fifth largest city, buzzing with the noise of engines driving mototaxis, air-conditioning units, and refrigeration devices. Its international airport is a major hub for Amazonian travel and trade, and despite further river movements the city has managed to maintain its status as a major Amazonian port. Of course, Iquitos does not escape the divisions, inequalities, and social injustices that characterize most large urban settlements. However, when we first visited Nauta in 2006 these problems mattered little to those we met. On the contrary, members of Nauta's established trading families in particular talked of their seventy-year longing for a fast terrestrial link to the city of Iquitos. This was the link that would sustain their sense of inclusion in a dynamic, open, and developing modern world.

Until the completion of the Iquitos-Nauta road the connection between the two urban centers was almost exclusively by river transport (see map 3).

A terrestrial route had nevertheless existed in the region since the 1950s, some say the 1940s. Local hunters always knew how to move about on land as well as by water, and a route of sorts was defined over the years. A short stretch of road leading out of Iquitos was constructed in the 1920s during Leguía's presidency, built under the auspices of the conscripted labor program that he initiated. President Belaúnde's declaration that rivers were the roads of the Amazon might have legitimately pertained to this part of Peru. However, river travel was relatively slow and extremely precarious, and it did not resonate with the sense of progress and national development so dear to Nauteños.

It was largely for this reason that a group of Nauteños took it upon themselves in the 1960s to make their own terrestrial link to Iquitos. It was a set of coincidences that led to us meeting the pioneer, Sr. José Domingo Murayari, who was responsible for leading the project to cut the first path through the jungle between Nauta and Iquitos. During a taxi ride along the newly built Iquitos-Nauta road, Hannah had mentioned our research to the taxi driver and much excitement had ensued. Leo, another passenger in the car, had said that he knew many people we should talk to in Nauta to find out about the history of the road, one of whom was a man called Murayari, the first pioneer. A hasty arrangement was made during the taxi ride that Leo would arrange a meeting with Murayari, and, sure enough, three days later Hannah was summoned to attend the meeting. Murayari, now an elderly man, had come down the river from a settlement some miles upstream especially for the meeting. When Hannah and Leo arrived at the house, Murayari sat surrounded by his family members ready to tell his story. The tale was compelling and yet poignant, capturing both the dangers and the personal sacrifices of the journey. In 1961, the mayor of Nauta, Señor Olivar, had approached Murayari and asked him to lead an expedition to discover a potential route for a new road between Nauta and Iquitos. He had heard about Murayari's skill in navigating the jungle, which, according to his family, who helped to tell the story, was legendary in the region. Murayari's form of navigation involved no technical instruments; it was based on his past experience of moving around the forest and relied on his knowledge of the stars, the wind, and the trees around him. To make the journey, Murayari had put together a team of nine men, including a cook, a porter, and the path cutters (*trocheros*). Beyond their machetes these men had no special clothing, just the boots on their feet and whatever the forest could offer them for comfort, such as leaves for making shelters at night and branches for fashioning themselves a platform for a bed.

The journey was long, hot, wet, and arduous and took twenty-one days to complete. On arrival in Iquitos, Murayari and the other pioneers were met by an engineer called Morana who drove them to the center of the city, where they were feted by local journalists, put up in a hotel, and asked repeatedly to recount what had happened. The journalists were not that interested in the hardships the pioneers had endured or in their feat of having made the first terrestrial link between Nauta and Iquitos. They wanted to know if there were commercially marketable hardwoods along the route and how suitable the land was for cultivation. From the outset, then, the project of constructing the Iquitos-Nauta road exceeded the promise of connectivity that had provided the impetus for this pioneering journey. Even at this early stage, the potential of the road was understood as much for its capacity to create new economic futures for city dwellers as it was to improve the social and economic fortunes of the towns it would connect.

The energetic extraction of natural resources was clearly associated with economic growth, and the potential of the Iquitos-Nauta road to turn a profit was recognized decades before the full stretch was eventually completed. The early opening of this territory attracted interest in the potential of both agricultural land and forest resources, and state officials continued to devolve powers to entrepreneurial traders and explorers in exchange for creating access to these territories. President Belaúnde himself was said to have made a deal with Raul Moreira, a prominent rubber trader, giving him the right to sell plots of land. These "sales" sparked the first set of lawsuits in which this one hundred kilometer stretch of land was to become embroiled.

Construction work did not extend much beyond the outskirts of Iquitos until the 1970s, when Augusto Freyre, a general in Juan Velasco's military government (1968–75), determined to make a proper road. Velasco's government was a radical regime, famous for the Agrarian Reform of 1969. The generals were high-profile public figures, responsible for the delivery of public works, which were often executed by the engineering divisions of the Peruvian army. Freyre was remembered in Iquitos as an elegant man who liked to parade his stylish girlfriends and act the part of the dashing urban cosmopolitan. But for all the trappings—which the Iquitos elites were able to deploy to demonstrate their links to a world of fashion and urban style—the isolation of this region also allowed these public figures to take advantage of crucial disconnections from central state controls. With respect to these early construction initiatives, the local representatives of the Ministry of Transport and Communications had little authority over the engineering battalions in the field. Thus, while the rationale that provided the funding for this first phase of construction combined a commitment to border security and to economic development, the intention of those on the ground seems to have been simply to make money. Among the people who talked to us about this period there was general agreement that the army systematically exploited the area and made little effort to actually build a road beyond the surface that facilitated their use of heavy machinery, and most of them assumed that it was connection to the supplies of valuable timber, not to the town of Nauta, that motivated this first phase of construction at the Iquitos end.

Almost as if to commemorate this turning away from a duty of public care, people name the point at which the road deviates from the direct route between Iquitos and Nauta as the deviation or detour (*el desvío*). In the 1970s a military camp was set up at the *desvío*, and no further attempts were made to continue on toward Nauta. The road led into the heart of what was, at that time, dense virgin forest. As one of the engineers responsible for finishing the road in 2006 explained, "The army just blasted through with their tractors. They pushed the earth to one side, creating uneven places where water got trapped. It has made the whole area much more difficult to work to a decent standard. But they were trained to a military mentality 'get in as fast as you can, kill the enemy, and that's it!'" (llegar lo más rapido possible, matar al enemigo, y punto!). The "enemy" at that time seems to have been the forest itself. There is a monument to the soldiers who worked on this stretch of the road at the place known simply as Kilometer 21. Referred to locally as the little soldier (*el soldadito*), the figure elicits an amused affection in people and some sympathy for the young conscripts who must have suffered as they opened up this area of virgin forest for profits that they were unlikely to partake in. The plundering is remembered, but so is the fact that even such plunder kept the road project alive.

Farther down the road, just five kilometers from Nauta itself, another statue was erected in memory of the Nauta pioneers, the civilian colonists who had campaigned to bring their needs and desires for a road to the government's attention; in that spirit, they had tried with machetes rather than machines to open the first routes out of Nauta toward Iquitos.

Figure 2. The monument of the "little soldier" at kilometer five

In 2005 this statue was gathering dust in a shed by the side of the road, waiting for this last five kilometers to be completed before it received a new coat of paint and a reinstallation ceremony. The two statues reveal a tension that is integral to the history of road building in Peru: between state-led construction programs, on the one hand, and local energies on the other. At times local people have felt that they needed to do everything in their power to keep their needs and desires visible to state agencies. At other times, it is clear that the state appeared via imposed policies that radically influenced how people

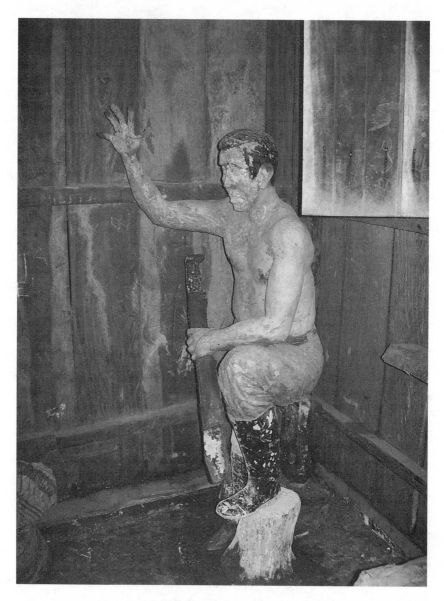

Figure 3. Statue of the first pioneer in Nauta

came to live in these arterial spaces, which were gradually opening up across the Peruvian national territory.

By the 1980s the fourteen or so families in Nauta who saw themselves as cosmopolitans had still not succeeded in their quest for a proper connection to Iquitos. These families, who tended to educate their children in Lima or even abroad in the United States or Europe, lived alongside the local indigenous population, much as the highland landlords had lived cheek by jowl with Andean peasant farmers. In these highly stratified communities, the elites showed little interest in the aspirations and interests of those they viewed as their social inferiors. Nevertheless the aspirations were strongly felt throughout the population, and the decades following the Agrarian Reform of 1969 saw increased investments in extractive industries (in this region particularly in timber and oil exploitation) and waves of migration supported by the central state, which encouraged poor Andean farmers to settle in the Amazon region. During the 1970s, Iquitos had seen a dramatic rise in its population as people came looking for work in oil exploration.[18] However, the labor-intensive exploration phase of the oil industry was soon over, and men who had been earning a wage by laboring in the forest found themselves back in Iquitos, stuck in a city with no land on which they could farm, few employment prospects, and large families to support. Meanwhile, the military government had relaxed international trade restrictions that had previously made commerce a viable industry in the area, thus limiting the possibilities of making a living in the commercial sphere.[19]

At the same time, institutional arrangements appeared that made the area that had been opened up by the military's efforts to extend the Iquitos-Nauta road suddenly look like a viable space of habitation. As part of a move to encourage small-scale entrepreneurial farming, the national agrarian bank (Banco Agraria) began offering loans to people who could prove that they had land to cultivate.[20] This attracted large numbers of colonists (*colonos*) to the road, who established settlements with names like Ex-Petroleros (Ex-Oilworkers), Nuevo Horizonte (New Horizon), and El Triunfo (The Triumph). These names indexed the histories that people brought with them and the hopes that imbued their future imaginaries.

This was the impetus that marked the next phase of the history of this road—and provoked the next great scandal of diverted resources. A Brazilian company that had been working in the nearby Napo area won the contract to continue construction. There were accusations of bribery and corruption, but local politicians argued that this (foreign) company had been able to make a more competitive bid than its Peruvian counterparts, as they already had their heavy machinery operational in the region. New controversies then arose around accusations of discrimination against local contractors and suggestions that politicians were simply awarding the contracts to those who paid them the highest bribe. These swirling speculations about shady profits and irregular contracts were matched by severe environmental consequences. Migrants arrived primarily from the Andean regions of San Martin and the riverine communities of western Amazonia. Deforestation followed the line of the road. By the 1980s, the malarial mosquito, which also followed the road, became endemic to the region, establishing breeding grounds in the swampy pools produced by the military's earlier incursions.

Progress on the construction went in fits and starts, and the connective potential of the Nauta road wavered as markets came and went, as the state appeared and disappeared, and as individual fortunes and enthusiasms waxed and waned. The decline of the national economy toward the end of Alan Garcia's first presidency (1985–90) put an end to the support for planned settlements, and squatters and resettled workers alike faced the problem of how to make a living in a region where the soils were not suitable for agriculture, productivity was low, and there was lack of support for generating sustainable alternatives. Some initiatives were abandoned, such as a mozzarella cheese factory that stood out as one of the less viable plans for the region. A processing plant had been built without serious consideration of how many water buffalo the local land could sustain. The small herd never produced enough milk, and by the time the plant was closed down it emerged that the managers had been flying in powdered milk from Lima just to keep the concern going. Other, more central public services, such as the basic educational needs of the area, were never fully met by the state either. In 2006 the key schools along the road were run by a French religious NGO, Fe y Alegria.

These rhythms of hope and disappointment orchestrated the ways in which people told us about how the road shaped local histories and carried the traces of how those histories unfolded. In the 1990s, with President Fujimori's more overtly neoliberal regime, a restructuring of the state produced the first experiments in devolved regional government. In this period the road again took on an aura of potential with the idea that it would help to create a stronger regional economy, offering a more stable sense of connectivity than the river alone would provide. In the early years of his presidency, Fujimori invested in public works. He built schools and fish farms, and he invested in roads. It was a period of state presence, but it was also a period of great political instability as civil war raged farther south in the Andean heartlands of Peru. After Fujimori staged an auto–coup d'etat in 1992 in order to suspend the Constitution and dismiss the judiciary in a bid to secure a third term of office, a series of different presidents of the Loreto regional government annulled and subsequently reinstated construction contracts. There was money flowing into the regional coffers at this time. Loreto had a privileged position in relation to the retention of oil revenues, and regional leaders also cut a deal with the Fujimori government at a time when he needed the army (which was tied up in border disputes with Ecuador) to fight the Shining Path "guerrillas" active in central Peru. Fujimori ultimately secured the finances to complete the Iquitos-Nauta road in exchange for the Loretanos's "acceptance" of a handover of contested Loretano territory to Ecuador. But again the promised sense of inclusion and connectivity was short lived. Much of the money for the road disappeared into private pockets, and although many people subsequently went to jail for embezzlement (as we discuss in more detail in chapter 5), the road did not quite make it to Nauta. Five kilometers remained unfinished, in dispute, with materials impounded and legal proceedings rapidly finishing off what was left of the budget. In 2000, President Fujimori fled the country under the cloud of serious charges of theft and corruption. The road was still not complete, and some sections had been built to such a low specification (the money having been spent on other things) that the surface was crumbling and virtually impassable in places. Nevertheless, despite the fact that a road in some respects is never exactly finished (especially one that has been over seventy years in the making), the final stretch was finally inaugurated in July 2005 by President Alejandro Toledo.

The social effects that this road has brought about over the years are complex and multiple. Both the construction work and the opportunities for the exploitation of forest resources—particularly land, hardwoods, and oil—have brought tens of thousands of people to the area. It is also clear that the road, and particularly its use by illegal loggers, has been very destructive of an extremely fragile ecosystem, and the stagnant water trapped by construction debris has allowed the malarial mosquito to take hold in the region. Despite these negative effects, road construction also holds an energetic promise that creates strong alliances across the social spectrum. At no point in our research conversations did any except the most committed environmentalists voice the opinion that the area would be better off without the road. On the contrary, local communities along the route have used their own efforts to connect and to reorient themselves away from the river to the road, sometimes building spur roads, sometimes relocating their houses. Living on the roadside has not as yet provided that sense of belonging and integration that people long for, but it is nevertheless generally understood as a step in the right direction. However powerful the river, the all-weather asphalt road is seen as more reliable, more stable, and less dangerous. People in Nauta are overjoyed that they now have a fast connection to Iquitos. A journey that previously took all night by river, and often several days by land when weather conditions turned parts of the route to deep, impassable mud, can now be reliably completed in under two hours, often much less. Nevertheless, this newfound connectivity produces new forms of marginality (as we discuss in more detail in chapter 2). In the short term it seems as if the Nauta road has left the town more isolated. Previously, for example, Nauta had hosted a high number of public employees—teachers in the various primary schools, medical staff from the local hospital, and other representatives of government agencies. However, the increased proximity to Iquitos has allowed this group of salaried consumers to live in Iquitos and to commute daily in the taxis and buses run by entrepreneurs from the city. They no longer spend their salaries in Nauta. The new hope is that the road will bring tourists; as we left the field, Nauta residents were thinking about how to turn the town into a tourist resort. They had built an observation tower at what is now promoted as the source of the Amazon, the point at which the Marañón meets the Ucayali. A concrete structure was erected at an embarkation point near the center of the town, although it was never staffed or furnished in any way. Similarly, a model fish market was constructed on the outskirts, in an area designated as the trade port—but again there was no local understanding of who was going to use this facility or when it might be brought to life. In 2006–07 the concrete structures seemed to signal the possibility of development, but also the danger that enthusiasm and concrete would not, in and of themselves, usher in the economic growth they had been hoping for. The road had by now become an active and dynamic space, with constant traffic and new businesses, and yet there were lingering concerns about its capacity to transform the future, a subject that we explore in more detail in chapter 2.

As we have demonstrated, infrastructures offer a powerful site for interrogating the waxing and waning of state politics. By following the histories of our two roads, we have been able to infer a general pattern that links these two particular roads to the wider socioeconomic conditions of the emergent Peruvian state, and indeed to broader international contexts in which these infrastructures were unfolding. In the late nineteenth century and early twentieth century the state was fragile, its borders were under constant threat, its coffers were empty, and the frontiers of state power were thus spaces where private enterprise, foreign companies, and entrepreneurial individuals shaped the national economy. The roads served these interests—but they were also integral to the possibilities for mobility and new beginnings for thousands of people. The ambition to unite the country through the territorial effects of infrastructural provision characterized a second phase of state formation, expressed most clearly in the Belaúnde years, in his two periods of office before and after the radical military regime of General Juan Velasco Alvarado. In the following chapters we turn our attention more explicitly to the third phase of road construction, which emerged after the war between the state and the Maoist-inspired guerrilla forces of the Shining Path. During this war, roadblocks were instituted by guerrilla and army personnel alike, and travel became extremely dangerous in many parts of the country. Since the war, successive presidents have favored neoliberal approaches to economic development. Infrastructural ambitions have focused less on integration and welfare and more on the possibilities of macroeconomic growth afforded by the mining sector, the development of Peru's ports, and the generation of hydropower. As we explore in later chapters, the current Interoceanic Highway project exemplifies a foreign investment from which the Peruvian state aims to reap financial benefit. Local populations are not key players in these political debates. And yet, as we follow in more detail in chapter 2, it is local people who shape the life of the roads, just as these roads are material forces that in turn shape the lives of those who live along and around them.

Chapter 2

INTEGRATION AND DIFFERENCE

Although infrastructural projects aim to effect their transformative potential through the promise of connective integration, the question of who or what is being integrated remains a matter of constant concern. Although both the roads we studied were generally conceived as initiatives to bring together previously disconnected places, expectations about what such connection entailed were far from settled. As we began to show in chapter 1, the integrative ambition of a new road must always contend with prior geographies and previous histories of connectivity. As Akhil Gupta and James Ferguson have argued, "If one begins with the premise that spaces have always been hierarchically interconnected, instead of naturally disconnected, then cultural and social change becomes not a matter of cultural contact and articulation but one of rethinking difference through connection" (Gupta and Ferguson 1992, 8). This focus on the differentiating effects of specific modes of connectivity is fundamental to the emergent politics of roads that we discussed in the introduction, and which this chapter traces in more detail by analyzing the narratives of the frontier that surround road construction projects. Our aim in this chapter is to show that while many people assumed change would result from the penetration of previously unexplored frontiers, each frontier in turn entailed its own complex geographies of connection and disconnection.

The road construction projects we followed took place in areas that were already inhabited, and explorations by those who assumed the status of pioneers routinely produced unexpected evidence of this prior connectivity. During our time with Don Benedicto Kalinowski he recounted a dramatic tale of one such encounter with evidence of past connection. He had been traveling in the jungle of Madre de Dios with the Dominican priest Padre Alvarez when they had come across a group of native people. The story began as a familiar frontier tale of discovery and encounter: Padre Alvarez and Benedicto managed an improvised communication that rested on mutual imitation, the offering of gifts (knives and mirrors for the men and clothes for the women), and gestures of friendship, alliance, and humility. Benedicto said that they had approached three tribes in this way and had won their trust. But the fourth was different. They came across these people as they were paddling downriver in their canoe. Standing there on the beach, these men had looked so terrible and so strange, "there they stood, bloodthirsty warriors, ready to kill anyone who could not escape their ravages!" He and the padre were armed, but they knew that violence was the most risky strategy, one likely to elicit a fatal counterattack. However, as they drew closer Benedicto was amazed to see that not only were they not brandishing weapons of any kind but they were standing there, deep in the heat and humidity of the forest, dressed as if from an Andean historical drama, in knee-length trousers and something that resembled a frock coat, complete with buttons down the front!

He told the story well, and we were captivated. What were Don Benedicto and Padre Alvarez to make of these strange people, apparent hybrids of highland and lowland, in the space that Michael Taussig has evocatively called a "landscape of the imagination" (Taussig 1987, 287)? Maybe these people had long since traveled down from the highlands to settle in this region. However, if their clothes could be taken as an indicator of their highland origins, their language could not, and Don Benedicto's attempts to talk to them in Quechua produced looks of confusion and incomprehension. The clothes, it seems, were a historical throwback to a previous encounter during colonial or even Inca times. Or perhaps they were a group of

Andean peasants, driven from their highland villages by political violence. He held us in suspense, rehearsing his ideas of how they might have lost their indigenous language but not, for some reason, their sartorial style. But he was holding back an extraordinary punch line. As Don Benedicto and Padre Alvarez got closer, they realized that the men standing on the beach were not in fact wearing any clothes at all. The trousers, waistcoats, and buttons were painted onto their bodies, with a care and detail that left him with questions and a sense of wonder that redefined his relationship to the forest in which he had lived all his life.

It was a great story—and one he must have rehearsed many times. In time he learned more about who they were, and abandoning all reference to his first impressions (or projections) of fear and ferocity, he told us that they were Amarakaeri people. Padre Alvarez was also moved by these strange, tall, peaceful people who he decided would be perfect subjects with whom to found the Christian mission of Shintuya, deep in a region that he clearly felt was in desperate need of salvation.

The power of Kalinowski's tale about his encounter with the Amarakaeri hinged on his shock at discovering an enduring connection between contemporary lowland Amazonia and the colonial highlands. This opaque reference to an enduring connection between Andean and Amazonian people beyond the better known trading routes and missionary incursions stands in stark contrast to more common renderings of this region of the Amazon forest as the home of peoples never before contacted by outsiders (Slater 2002).[1] If Gupta and Ferguson encourage us to be attentive to rethinking difference through connection, Kalinowski's story reminds us that not all histories of connection are equally legible. This reminder is instructive if we are to understand the ways in which infrastructural transformations involve differential possibilities for making claims about who should be integrated, and what they should be integrated into.

We spoke with many people about their knowledge and recollections of the histories of both the Iquitos-Nauta road and the Interoceanic Highway. In these conversations the trope of the frontier was routinely deployed to conjure a sense of incursion into the unknown. The unknowns of the frontier are marked by specific absences, notably the absence of civilization, state controls, and modern life. Thus, even where previous life worlds were encountered, known about, or imagined, they were worlds that lacked the qualities that the new infrastructures promised to introduce. Most notably, they cast so-called frontier regions as empty places that are nevertheless full of generative potential for newcomers to make of them what they wish. Attempts to fulfill the promise of the frontier thus inevitably produce conflictual or disturbing encounters precisely because preexisting life worlds are rendered invisible by the myth of the frontier. Far from being an empty space, however, we suggest the frontier is better characterized as what Anna Tsing has called a "zone of awkward engagement" (Tsing 2004), where dehistoricized autochthonous cultural and natural relations become part of a politics of differentiation through which future possibilities associated with the frontier are framed. It is to the modes of differentiation that we found at play on both the Interoceanic Highway and the Iquitos-Nauta road that this chapter turns.

On the Iquitos-Nauta Road

One of the key discussions that took place along the Iquitos-Nauta road was a concern with ownership of the land that had been made available by the building of the highway. The first people to claim ownership of the land along the road were the army generals who were involved in the road construction process itself.[2] As the road construction proceeded, land that had previously been inaccessible came to be seen as a potential investment, and several army generals took the opportunity to stake a claim to plots. Many told us of the underhanded methods by which these generals managed to get their hands on land titles.[3] The travesty of these land claims was exacerbated by the fact that, of those generals that claimed the land, none of them actually chose to live on the plots they owned. Instead, they lived in residences in the nearby city of Iquitos or in the capital of Lima, while the land along the road was left fallow, lying in wait for the day when the road was finished and its latent economic value might be realized.

Although for these generals the value of the land lay in its speculative potential, there were others for whom this legally owned, but ostensibly abandoned, land simply looked static and unused. For many people who were looking for somewhere to make a livelihood, the abandoned plots looked full of more immediate potential for agricultural development.

Nelson was a member of one of ten families who, frustrated with their lives in Iquitos, decided to group together and to travel out along the new road in search of what he called a "new horizon" (*nuevo horizonte*). Their aim was to find unoccupied land that they might live on, or at least that they might cultivate while their families continued to live in Iquitos. Nelson and his companions had approached the Ministry of Agriculture in the hope of finding out how they could gain use-rights for land that appeared to have been abandoned. The Ministry of Agriculture was under considerable pressure at this time to assign lots to these new settlers, and it had a record of overturning earlier claims that the generals had made on the land. However, Nelson recounted how they were confronted with a ministry that was used to dealing in underhanded ways with land claims and which had no coherent land settlement plans or strategies for the Iquitos-Nauta road. He told us that they were advised by the ministry that the only way in which they would be able to gain any kind of ownership rights to the land would be to occupy a plot. Following this advice, they set out on an expedition along the road to the point at which the current phase of construction had stalled on the banks of the Itaya River. At this stage, the paved part of the road proceeded until kilometer twenty-five, after which the terrain was sticky and muddy and only traversable on foot. From the end of the paved road, the group had walked another thirty-one kilometers over rough terrain to arrive at their destination.

After a journey of nearly twenty-four hours, they arrived at the Rio Itaya, but to their surprise the land they came across was not empty, as they had expected. On the banks of the river they found another nascent community of people who had built their houses there, with the land already parceled out into agricultural plots. Undeterred by this discovery, they backtracked toward Iquitos, where they found another place that they decided they might be able to settle, on the banks of a stream at km 26. Here they set down their belongings and began to work the land.

It turned out that this plot of land was already owned by an army general. Nelson told us how the general confronted the group of colonists, but the colonists held their ground, and managed to maintain the right to inhabit the land themselves. On July 24, 1985, documents were signed that marked the foundation of the Agricultural Association of Paujil. Some 350 people followed these original colonists to become part of this association, and by 1989 the community had gained sufficient recognition to secure funding from the regional government for a four-kilometer track leading from the main road into the forest, thus extending considerably the land available to the association.

A former army officer told us that the colonizers had been responsible for stealing over two thousand square miles from the generals who had first claimed the land, but that there was nothing the generals could really do about it. Ownership is weak when land is unoccupied, and along the Iquitos-Nauta road occupation appeared to trump the threat of force when it came to winning land rights. Against a background of national land reform, which was under way at the time, it is perhaps unsurprising that the associations found it relatively easy to displace the generals as individual landowners. Moreover, the 1970s and 1980s had seen an increase in labor organization in Iquitos. By the time the road was being colonized, many of the unions were powerful voices in the region, having brought a much greater sense of community activism and local rights to the roadside dwellers (Rodriguez 1991).

Nonetheless, the generals were not going to go quietly; they had a reputation for making life as difficult as they could for the colonists. At the same time as agrarian associations like Paujil were being established, the army placed a barrier on the road at kilometer twenty-five beyond which they would not allow any vehicles to pass. They argued that beyond this point the road was too poor for vehicles to travel on. Meanwhile, they apparently made no effort to improve the quality of the road, in spite of the fact that during this time they were officially responsible for its construction and maintenance. Colonists were thus forced to carry their tools, belongings, and products on their shoulders, often for many kilometers, to the places where they were living.[4]

As the road opened up access to new land, an initial interest in the economic potential of this space was enacted in a range of particular actions, initiatives, and arrangements, which themselves had consequences. Ambiguities over occupation and ownership were played out in the tensions between the speculative ambitions of army generals and the desire of colonists to set up small farms. Meanwhile, the increase in the number of people who were now living on individual landholdings along the road was beginning to lead to attempts to consolidate some of these distributed clusters of families into more coherent and integrated communities.

The members of the Agricultural Association of Paujil were initially drawn to the area because of the possibilities it afforded them in terms of productive activities. The space allowed them to establish fish farms and grow fruit and vegetables, which they could sell in the markets in Iquitos; it even enabled them to create a retreat away from the hubbub of the city. As colonizers built houses and became more established, however, attention began to shift from the possibilities that the settlements afforded individual families to the possibility of transforming what was for a long time a loose network of landowners into a more formal community, focused around plans for a community center, health post, and school. Concerns about economic productivity were thus soon translated into a preoccupation with social integration through the establishment of community.[5]

The River and the Road

At the same time as these original pioneer families began to try to consolidate themselves as community organizations, the completion of the road prompted the arrival of groups of people who from the outset attempted to establish their own roadside communities. Since the construction of the highway, many new villages have sprung up. These settlements are often named after the date when they achieved formal recognition as settlements. People living in these villages often come from similar parts of the country and are keen to engage in both economically and socially productive activities. Many make a living by cultivating plots of land, where, in spite of the reputedly bad soil, they use slash and burn techniques to clear the land on which they grow crops such as yucca, pineapple, maize, and banana. Others produce charcoal and firewood to serve industries such as brick manufacture, and some set up small shops and bars to serve neighbors and travelers on the road. Meanwhile, villagers in these settlements have been active in petitioning local governments for investments in their settlements, including grants for community centers and schools.

These villages are inhabited both by people who come from towns that are distantly connected to Nauta and Iquitos via the Peruvian river system, such as Contamana, Yurimaguas, Tarapoto, Pucallpa, and San Martín, and by people who have moved from smaller villages upriver and down. Whereas the former have traveled to the region for a variety of personal and economic reasons, the latter have primarily come in search of markets where they can buy and sell produce and to be closer to hospitals and schools. The riverside villages in this part of the Peruvian Amazon have a reputation for being difficult places to live. Malaria and other tropical diseases are a big problem, health care is very scarce, and opportunities for commerce are limited. The decision to move closer to the road is therefore provoked as much by changing

circumstances in the places that people have come from as it is by the transformative potential of the road itself.

The current problems that people face in riverside villages have a long history. Many of these villages were established when the rubber industry collapsed in the early twentieth century. People who had been entangled in in-
dentured labor and the slavery of rubber exploration found themselves living
on abandoned rubber plantations on the edges of rivers. After the rubber barons left, they organized themselves into communities and engaged in a combination of hunting and gathering, subsistence agriculture, and small-scale
exploitation of natural resources. Although free from the contracts of the rubber barons, the river traders (*regatones*) continued their relation of dependence
on outsiders. They were drawn into systems of uneven exchange, providing
products such as wood, *barbasco* (a product used in the making of insecti-
cide and fish poison), and animal pelts that could fetch high prices on international markets, in exchange for basic commodities such as soap, salt, and
clothes bought on account.[6] The liberalization of trade during the 1980s, and
a fall in demand for products due to the development of synthetic alternatives, led to the decline of even this commerce, making lives in these riverside communities very difficult. As the river has waned in importance as a
space of economic flow, the opportunities offered by the Iquitos-Nauta road
have in many ways mirrored the kinds of opportunities that had previously
flourished along the rivers.

Some people pointed out to us that the road thus offers a similar structural role in these settlers' lives as the river did in previous times. As the road
began to replace the river as the conduit for people and goods, some of the
settlements, which were previously concentrated on the banks of the nearby
Itaya and Nanay Rivers, began to move away from the river as people increasingly moved to live nearer to the road. The village of Santa Marta, for example, which is located on the edge of the Nanay River about five kilometers back from the road, responded to the building of the highway by
cutting a two-mile path through the forest from their village to the road,
giving the villagers quick access to hospitals, schools, and transportation
into Iquitos. In 2005, around half of the families in the community of Santa
Marta had moved up to live alongside the road, while a few families stayed
down by the river where their land was located.[7]

The Limits of Integration

Alongside the emergent pioneer communities, and the new villages just described, there are several plots along the road that are still owned by absent landlords. To ensure that the land does not get invaded, landlords have employed guards to look after the land. Unlike the colonists, the guards do not make any kind of claim to permanent residence on the land. Rather, absentee landlords pay them a small stipend in return for their presence and their protection. Unlike the other people who live along the road, the guards do not appear to enter into the spirit of the frontier economy, either in terms of engaging in the economic exploitation of the land or local markets or participating in any kind of social or community formation. Unlike the settlers, who are engaged in small-scale agriculture or commerce, the guards are not allowed to cultivate the land that they are living on.

Many of those living along the road therefore see the guards as the people who have it the hardest. Although they are paid a wage for their protection services, they do not seem able to make anything of the land they are living on. In this sense, they seem to be anomalies in this frontier space. However, there is another story of who the guards are, and how they came to be here, that provides an alternative explanation for why they would have chosen to work as guards rather than occupying the land and turning it to their own economically and socially productive advantage. The guards are, like the original residents of Nauta, Cocama people.

To describe the guards as Cocama is, for many people, to provide a reason why they would be happy living on land that they have no capacity to make a claim of ownership on, which they do not cultivate, and in which they have no social ties. Cocama are described as being able to live on the estates through a traditional life of hunting and gathering, indexing for many a link to a precolonial or preindustrial relationship that Cocama people have to the land. They are understood to work as guards as long as the land they are looking after continues to provide for them. If and when it becomes exhausted, it is expected that they will return to their communities, which are located farther into the forest.

One effect of the presence of these Cocama guards in the space of the Iquitos-Nauta road has been to draw people's attention to the way in which the current settlement of land has not been a process of colonizing a socially empty space, but was rather an instance of inhabiting space that was already previously inhabited by Cocama people, who related to the land through movement, hunting, and gathering. The politics of using land that might otherwise have been used by indigenous groups engaged in traditional subsistence practices is further complicated by the historical marginalization of Cocama peoples in this part of Peru. The term "Cocama" as a label of ethnic identification has over time become increasingly problematic.[8] The Cocama were the original founders of the Nauta settlement. But nowadays to call someone Cocama is more often a term of abuse than a simple description of a particular cultural identity. Given this, those who continue to self-identify as Cocama have become socially invisible, a situation that is now being responded to through the implementation of a bilingual and cultural education program that aims to reintroduce Cocama practices and customs to communities where these practices are at risk of being forgotten.[9]

Given the social invisibility of Cocama customs and practices, it is striking that the nonextractive relationship that the guards have to the land provides a putative link to an otherwise undetectable Cocama heritage. The attribution of Cocama identity to these guards appears to explain why they would act as guards. Their current employment is interpreted by other settlers as linking them to a historically erased cultural identity that is only now beginning to be recovered through programs in cultural education and the promotion of bilingual education in the region.

We return to the confrontation between indigenous groups and the activities of other social groups living along these roads in our discussion of the Interoceanic Highway. For now, we conclude our description of some of the dynamics of social differentiation on the Iquitos-Nauta road with a consideration of a final actor that we have not yet considered—the land itself.

If the presence of the Cocama guards in part points to a set of usually unacknowledged relationships that existed prior to more recent land invasions, appeals to the preservation of the natural environment provide another.[10] Environmental concerns about the impact of roads on the ecosystems through which they pass have been particularly prominent in the Amazon region, where processes of deforestation have been directly linked to road-building programs and the activities of the colonists who have migrated to their environs.[11] On the Iquitos-Nauta road, the extractive and agricultural activities we have described occur perilously close to fragile "white sand" forests on the banks of the Nanay River, which came under official protection in 1999 with the setting up of the Zona Reservada Allpahuayo-Mishana. Moreover, as Pepe Alvarez, a charismatic Spanish biologist and environmental campaigner who works in the Peruvian Amazon Research Institute in Iquitos, told us, the building of the road has led to dramatic deforestation, which threatens to turn fragile and highly biodiverse local ecosystems into very poor quality agricultural land.

Alvarez was particularly concerned about biodiversity. He told us that the area that the road passes through is one of the most biodiverse regions in the world, with over three hundred bird species and 142 reptile species detectable within a hectare. In 2005, Alvarez was responsible for identifying a particular bird species, the perla, as being under threat of extinction in the region, and his work has been concerned with finding ways of creating nature reserves to protect the fragile biodiversity from the kinds of colonization practices that we have described above.

Although Alvarez was concerned about species biodiversity, he was also at pains to point out that this did not mean he was against development. "People say I care more about birds than poor people," he told us, "but destroying biodiversity does not help poor people, it only benefits the rich, and only in the short term." Alvarez's concerns appeared to stem not from a simple relationship, like that which Teresa Brennan (2000) describes, between capitalist expansion and the exhaustion of the environment, with the assumption that curbing the former will solve the latter. Rather, conservationists like Alvarez found themselves confronting the challenge of how to make a claim for the continuation of what are taken as ahistorical natural relationships—in the form of national parks, reserves, or practices of protection—that seem to stand in opposition to the realization of progressive futures. In this context, the notion of the frontier as standing for disconnected "nature" anchors these fragile environments in a stagnant past. Previous relations become that which has to be overcome or displaced for development to succeed, and for more prosperous futures arrived at.

If the history of the Iquitos-Nauta road can be understood in one sense as a manifestation of a dream of prosperity through connectivity, which we have attempted to draw attention to in the first half of this chapter, the road has also provoked an unfolding of social relations that are not determined by—but still follow from—the ambitions for progress that we discussed in chapter 1. On the Iquitos-Nauta road, people are not only preoccupied and concerned about the connective possibilities of the new infrastructure. They also have questions about appropriate ways of facing the future through, for example, speculation versus cultivation; the importance of community formation in spaces of uncertain or as yet unformed sociality; and the unsettled tension between the value of cultural/natural heritage and the promise of transformation. Although this is but a cursory look at the complex and emergent politics of differentiation that we found on the Iquitos-Nauta road, an awareness of both the power and the contingency of these relational dynamics provides an important basis from which to understand the practices of the engineering consortium that we will explore in part II of the book. Before that, however, we need to trace some of the specific ways in which these tensions played out on the Interoceanic Highway.

Life on the Interoceanic Highway

As we recounted in chapter 1, Route 26 was completed long before the Iquitos-Nauta road. Since the 1980s, a road of sorts has existed along the whole length of the route that was to be paved as part of the Interoceanic Highway project. If our concern in relation to the Iquitos-Nauta road was the social relations that were produced by the possibility of opening up a new route through a forest, with all the economic and social arrangements this subsequently entailed, our interest in the emergent politics of the Interoceanic Highway is more focused on the way in which decades of migration along the highway has acted to constitute the finely differentiated sociality of this space.

As we described in the previous chapter, the history of the Interoceanic Highway has been characterized by several waves of migration, which has linked this region both to other areas of Peru and to the international circulation of people and capital. Much of the migration has been internal to Peru, with large numbers of people traveling from highland communities to work in logging and mining or as auxiliary workers to these industries for discrete periods of time before returning home to highland villages. Others based in the highlands have gained employment as drivers for firms transporting gasoline, wood, and commercial goods up and down the highway.

One of the striking effects of migration from the Andes to the Madre de Dios region has been the emergence of a notably Andean feel to these lowland territories. The difficulty of travel along the road might be held responsible for much of the continued entanglement of highland lives in the lives of those living in the lowland. We were often told how vehicles would break down en route, requiring assistance from fellow drivers, and how people found themselves stranded for days waiting for landslides to be cleared or flood waters to recede. They would have to find lodging in the local hotels and restaurants that dotted the route, where they would invariably bump into extended family members and in-laws who were either working on the roads or had previously moved to one of the lowland settlements. One time when we were traveling on the road near Puerto Maldonado the jeep we were in ran low on gasoline. Our drivers, Carlos and Tomás, were delighted when they saw a fuel truck (*cisterna*) that they recognized at a roadside café. Sitting in the front of the car, Carlos signaled to the driver, who was having a break in the restaurant, and managed to get his attention. He interrupted his meal to find out what Carlos wanted. Carlos explained our predicament and asked the driver if he would mind siphoning off a bit of his fuel just to get us to Puerto Maldonado. It turned out that Carlos was a distant relative of this man. Once this connection was established, the driver was very accommodating, and we were soon on our way.

Truck drivers, travelers, and tradesman along this route are not the lonely and isolated commuters of an anonymous highway, but are entangled in ongoing relations of kinship and friendship that connect the space of the road back to the places that it aims to join up. As well as relating to other drivers who ply the route, travelers often have relationships with people living in more permanent settlements that have sprung up around the extractive centers of logging and gold mining. When we arrived at the gold mining town of Laberinto, deep in the jungle region of Madre de Dios, we found the town peppered with signs and references to festivals, pilgrimages, and towns in the Andes, indicating the ongoing ties that people have to lives in the highland.

When we got to Puerto Maldonado we visited Lorena, a friend of Penny's, who moved from Ocongate in the Andean highlands to Puerto Maldonado several years ago. The family lives in a concrete house in one of the red-light districts of the city, where they run a small convenience store and a bar/brothel known as a "prostibar." Lorena said she had never wanted to run a prostibar, but all the canteens in this area worked in this way, and so she had no choice. It seemed a long way from the rural lifestyle of Ocongate. And yet in the yard behind the house we found a sheep that a relative had brought down to Puerto Maldonado from Ocongate a few months earlier. Lorena's niece was also visiting from Ocongate and is considering whether to stay in the city for her education.

Puerto Maldonado, like other towns along the road, is inhabited by many families like this who continue to have strong ties to the highlands. However, as we spent more time on the road, it became apparent that the Andean character of this space is complicated by a differentiation between those who have come from distinct parts of Peru. A particularly powerful distinction is made between those from the Cusco region and those who have migrated from the neighbouring region of Puno.

Halfway along the route of the road, at a town called Inambari, the highway splits. One branch of the road, which was the part we were studying, wends its way through the highlands to Cusco. The other branch leads to Puno. As Cusqueños, our drivers Tomás and Carlos were particularly suspicious about the places on the road that were inhabited by people from Puno. When we passed the village of San Lorenzo, located near where the road forks, they remarked that the place is known locally as "the town that God forgot." They said that the residents were engaged in illegal logging. However, it was not the illegality per se that prompted the negative epithet, but rather their reputation as people who lacked solidarity and denounced one another to the authorities rather than sticking together. Carlos and Tomás had no doubt that they behaved like this because they were from Puno. They also enjoyed telling us horrific stories of murder and betrayal in which the villains were inevitably Puneños.

Speaking to people living in the settlements along the road, it was clear that the juxtaposition of people from different places was central to the characterization of the area's reputation for danger, violence, illegality, and radical uncertainty. Questions of origin, displacement, anonymity, and living outside the law coursed through stories that we were told of people's experiences of living along the road. Issues associated with the impact of migration were particularly vividly portrayed when we had the opportunity to speak to a woman called Hilda, who lived in a community just outside the town of Mazuko. Hilda drew a clear distinction between permanent residents (*permanentes*) and outsiders (*los de afuera*).

She identified herself as a *permanente*, one of the few remaining members of a group of Amarakaeri speakers who had lived in this area for several generations. She told us how her forefathers had long inhabited the land, and explained the process by which they had achieved land titles in 1976–77. The plot she lives on extends from the road on one side, back about thirteen hundred meters to a river. However, Hilda told us that the land titles were proving to offer little protection against threats from outsiders who had come to the region to make money and mine gold, and who had few qualms about invading her land to do so.

Hilda reflected on the differences between the current migrants, brought to the area through the prospect of gold mining, and prior waves of migration. In the past, she told us, there were always people from lots of different places—Poles, Japanese, and people from the coast and the highlands. She recalled how earlier waves of migration had led to conflicts in her family, as people started to marry outside her own cultural group. Her mother's first husband had quite fair coloring (*medio-gringito*), which had led to a great deal of suspicion and envy (*evidia*) among her relatives. Now, however, there are so few indigenous people living in the vicinity of the road that the influx of outsiders is a real threat. She told us of one community where, due to in-marrying, there was just one elderly man who was truly indigenous. He recently died, and now the whole community is claiming indigenous rights even though none of them has any real claim to indigenous heritage.

Even in the indigenous communities alongside the road, then, claims to cultural identity are often not quite what they seem. Originary communities are as likely to be constituted by outsiders as migrant communities, unsettling the boundaries between *permanentes* and *los de afuera* in ways that make it difficult to know who or what to trust. During our conversation with Hilda, we took a photo of her, and we returned a few days later with a printout of the image. When we returned, there was a problem with the generator, which she said she bought from some Korean traders. This prompted Carlos and Tomás to launch into a tirade against the dangers of dealing with the Chinese, whom they saw as untrustworthy business partners. They cast Chinese businessmen as con men who would sell you an item, disappear with your money, and leave you to deal with the consequences when the true owner of the object came knocking at the door waving his proof of ownership in your face. Like all outsiders, the Chinese were described by our drivers as untrustworthy individuals, with their capacity for accumulation being put down to deviance and trickery, rather than some more legitimate form of trade relationship.

In frontier spaces like this, appearances can be deceptive. When people are obviously outsiders they raise suspicion—where have they come from, where does their wealth come from, and where are they going? Even apparent similarity can hide a multitude of differences. Carlos and Tomás delighted in telling us about a millionaire farmer in the Andes who still chooses to live in a hut on a bleak mountainside with his herd of alpacas. They also told us about a woman in Puno who made her fortune through smuggling contraband. Although she has a four-wheel-drive car, when you see her climbing out of it she is still dressed in her traditional homespun skirts (*polleras*), her money secreted away. With no stable means of reading who people are, where they come from, and what they have made of their lives, people living on the Interoceanic Highway have become experts at looking for the often uncanny effects and unlikely causes of differentiation in the landscape of the frontier economy.

Spectacular Wealth

Much of the migration that has taken place along the Interoceanic Highway has been prompted by the promise of fortunes that might be made from extractive activities on the frontier. We heard many stories of previous instances of the spectacular appearance and disappearance of wealth at earlier moments of expansion on the Interoceanic Highway. In Quince Mil, we were told about a time before the road arrived when gold was in such abundance that parties would be thrown where gold dust would be used as confetti. Nuggets of gold found in the rivers at this time were reported to have been the size of a man's fist. However, the price the miners had to pay for the potential access to these riches was living in lawless and dangerous gold mining communities. Life in these towns was deemed hard and lonely, and many miners were reported to have ended up squandering their fortunes on beer and prostitutes. Although the local market was flooded with gold, the lack of a road meant that the prices of commodities such as fuel and beer were hugely inflated. In 1960 the price of beer in Cusco was five soles, while in Puerto Maldonado you would have to pay double that. The difference in the price of petrol was even greater: four soles per gallon in Cusco, compared to eighteen soles per gallon in Puerto Maldonado.[12]

This combination of the abundance of gold and paucity of goods to spend it on had led some, we were told, into a state of insanity. Don Braulio, who we met in Quince Mil, recounted the story of a man who had come to be known locally as the "Human Beast" (Bestia Humana). Bestia Humana was a gold prospector who lived in Quince Mil. Originally from the town of Abancay in the department of Apurimac, he traveled frequently to Cusco, where he had a large network of important friends and associates with whom he conducted his affairs in a highly reputable manner. However, when he was down in the jungle, all façade of civility was stripped away, along with his clothes, and he would walk around naked, urinating in front of the women. He had a terrible reputation for torture and the murder of his fellow men in his dogged pursuit of riches from gold, which he then frittered away through excessive consumption. With no road, towns like Quince Mil were cut off and had no police force or law courts, so there were no mechanisms for putting a check on the horrific behavior of certain individuals who had gone mad in the face of the excesses of the gold economy.[13] Without some integration into state mechanisms, there was a constant threat of social collapse.

With the arrival of the road construction project, some of the dangers and anxieties around the appearance of spectacular wealth have begun to appear again. On the one hand, the appearance of the road has coincided with the 2008 crash in global financial markets, which has led to a rise in the price of gold, and to another gold rush. On the other hand, the very presence of the road construction process itself has produced a similarly unfathomable experience of spectacular accumulation destabilizing categories of social relations, to which we now turn our attention.

When the construction of the Interoceanic Highway began in earnest in 2006, large billboards celebrating the construction of the new road appeared at visible points along the way. In the Andes, they were adorned with a background picture of the archaeological site of Machu Picchu, onto which was pasted a photograph of two Andean women wearing traditional straw hats and brightly colored shawls around their shoulders.

In the Amazon the billboards showed an image of the mighty Amazon River surrounded by rainforests, with tropical birds and fronted by two smiling native children whose shiny costumes and laughing eyes made them look as if they had put on fancy dress for the photo shoot. Superimposed onto both backgrounds was a photograph of a large white Scania truck cruising along a dark tarmac strip scored down the middle by a bold painted double line.

What struck us first about these billboards was the depiction of the social spaces through which the roads were passing. The complex sociality of the route of the road that we have described thus far was completely ignored, substituted by a highly normative and traditional image of Andean and Amazonian life. At the same time, our attention was also drawn to the image of

Figure 4. Billboard advertising the Interoceanic Highway project in the Andes

Figure 5. Billboard advertising the Interoceanic Highway project in the lowlands

the juggernaut on the newly paved road. There was something about the juxtaposition of this symbol of universalizing progress, set uncannily against an incongruous background of other kinds of lives, that resonated with people's descriptions of the sudden appearance in their communities of the construction consortium with all its paraphernalia.

When these signs were erected, there were as yet no Scania trucks plying the route, but there were legions of powerful four-wheel-drive vehicles owned by the regional government and the engineering company that were already taxiing convoys of managers, engineers, and politicians to strategic meetings in towns along the road. As the road construction process progressed, these first symbols of capital expenditure in the region expanded into large fenced-off camps to house the managers and workers; teams of environmental scientists with GPS devices, cameras, and monitoring technologies; and huge pieces of road-building machinery. In this technical apparatus there appeared a display of ownership and wealth in a place that, despite decades of small-scale entrepreneurial activity and labor migration, has experienced relative marginality in terms of the circulation of commodities. What was also strik-ing was the way in which the appearance of these accoutrements of moder-nity remained cut off from the daily lives of the people living in the region through which the road passed. Although multiple state and corporate tech-nologies and personnel were now entering the frontier space of the road construction—the Internet, mobile telephony, security personnel, and state officials—their presence remained largely separated from most of the people who lived on the road. Drivers of the four-by-fours, for example, had been told not to give lifts to local people, who often tried to hitch a ride on the back of the vehicles. The camps that housed the road construction workers were built with tall fences, security cameras, and guard posts where anyone entering had to show their identification or passports to be able to go inside.

Inside the engineering camps, the distinctions between different kinds of workers were maintained by strict management of access to different mate-rial comforts. Workers slept in dormitories with bunks, ate in a separate can-teen, and were transported in buses. Middle-level managers were housed in twin rooms, had a comfortable communal area with television, ate different food in a different canteen, and traveled in private cars. The top managers had individual rooms, Internet access, a more salubrious lounge in which to relax, and were entitled to sit in the front seat of the four-by-four vehicles. Although these divisions are perhaps not surprising, they effectively empha-sized the channeling of symbols of capital accumulation toward *los de afuera* (those from elsewhere)—in this case, the managers from Brazil and Peru. Local residents more often than not found themselves on the other side of the walls of the engineering camps, from where it seemed it was only luck that would give them the opportunity of a temporary job on the construc-tion project and access to some of the commodities and materials that this would generate.

On one occasion, one of the drivers employed by the engineering consor-tium began to reflect on the wealth of tourists as another class of outsider, by asking himself why tourists in the region were so interested in taking photo-graphs of the mountains. He suggested that they actually had metal detec-tors in their cameras, and that they were not really taking photos at all but were surreptitiously prospecting for gold. When Penny suggested that she thought they were probably just tourists and were not interested in gold pros-pecting, he conceded that maybe only 50 percent were interested in looking for gold. Similar accusations were made of road engineers who people suspected were analyzing soil samples in pursuit of gold. Searching for explanations for how visible wealth often seemed to accrue indiscriminately was a common preoccupation for people confronted by the appearance of tourists, prospectors, and now the engineering consortium.

Engineers working for the construction consortium were not unaware of the disjunctive social effects produced by the appearance of a large-scale engineering project like the Interoceanic Highway project. In the construction camp near the town of Marcapata, the engineering consortium had put considerable money into making a comfortable communal building for the managers. The building came to be called "el club," replete with bar, TV, fireplaces, and polished wooden floors. It had been erected on rented land, and there was some concern over what would happen to the building when it was handed back to the landowner. The engineers wondered if any use would be found for it, or whether it would simply be demolished. In light of the ambiguous status of the club, and its juxtaposition to the otherwise rural and relatively poor town of Marcapata, a mural had been painted on the club wall that addressed an imaginary archaeologist from the future, who might excavate the material traces or the building, and realise that these were the remains of the strange nomadic tribe of the CONIRSAs. The mural read:

Archaeological Find

Recent excavations of archaeological remains on the banks of the river Araza on the western slopes of the southern Peruvian Andes have allowed scientists to claim, with great certainty, the discovery of the third settlement of the civilization "CONIRSA," in which they have uncovered a very distinctive and salubrious construction called "el club." Previous settlements were found on the banks of the rivers Ccatca and Mapucho.[14]

The CONIRSAs were a nomadic people, who inhabited this region approximately 1000 years ago, and whose obsession with constructing what they called "roads," "bridges," and "clubs," has surprised the intergalactic scientific community.

This new settlement, which may have been called Chiare or Limacpunco, presents, however, a much more careful and elaborate development. It was constructed in the twenty first century between 2007 and 2008 AD, by men and women who wore a piece of clothing called a "casco" [hard hat].

The CONIRSAs were permanently mobile, constructing without stopping. They used yellow machines and vehicles (that apparently had some significance that scientists have yet to uncover). The CONIRSAs communicated between themselves in very unique dialects which were created in the course of their work, and that came to be known as "portañol"[15] and "quechañol,"[16] inherited from old Brazilian and Portuguese travelers.

The roads were used in these times as a way of transporting terrestrial vehicles which used petrol and its derivatives. These disappeared more than 500 years ago. To traverse rivers they used structures called "bridges," the roads were very useful as a way of communicating between villages and allowed their development and well-being.

And the clubs . . . one can only imagine what the intended use was for a building made with such dedication and force!

Planet Earth, 20 August, 3030

The engineers working for the construction consortium were well aware of the disjunctive effects that their presence in the region was producing. Entering into a space that has itself been characterized by previous waves of migration, the engineering consortium might be seen to be just the latest in an ongoing movement of outsiders into the region that has been facilitated by the construction of roads. People frequently expressed anxieties in the towns along the Interoceanic Highway route about the implications of the new migrations that the road construction process would provoke. Aware of the social impacts of gold mining, people raised questions about the implications of an influx of mainly male workers into the region that would accompany the road construction project. There was much talk of prostitution and AIDS and of road-construction workers siring children with local women and then leaving them abandoned.

If contemporary road construction in some senses mirrors previous waves of migration in its stimulation of an influx of outsiders into the region, and in the part it plays in conjuring unfathomable juxtapositions of wealth and poverty, it also entails dynamics that are rather different from the contours of other kinds of unregulated migration. First, the construction consortium's awareness that it is entering a social space where it wishes to make a positive difference means that construction engineers arrive in the region with a different model of social transformation than that evoked by other migrants. Unlike the migrants who travel to work in extractive industries with the aim of earning money to take back to their families, or those who we met on the Iquitos-Nauta road who saw in the road an opportunity to rework and reinvent both their economic fortunes and the possibility of forging new communities, road engineering operates with a separation between the personal ambitions of those employed in the construction process and the general ambition for social transformation that it is hoped that the roads themselves will effect. As we explore in chapter 6, engineering consortia not only have an incidental relationship to the social worlds that confront them, but they have also taken on some degree of responsibility for improving the livelihoods they find there by integrating local places into the road construction process. This includes a commitment to employing local labourers, using local materials, sourcing food from local suppliers, and using local restaurants. The hope is that by stimulating a local economy in this way road construction might kickstart a development process that will allow people living alongside the new road to gain some of the benefits it appears to promise. This promise is, however, continually set alongside the visible social and economic divisions that the construction project participates in producing.

By drawing attention in this chapter to the various integrative promises and distintegrative threats associated with the two roads we have been studying, our aim has been to complicate our understanding of the politics of differentiation that inheres in road construction programs and in infrastructure projects more generally. Although roads are powerful technologies for achieving state projects of territorial integration, they are also spaces within which people are forced to grapple with ongoing questions about what kind of future they are engaged in, who this future is for, and how it should be brought about. On the Iquitos-Nauta road, these questions have largely circled around issues of land rights, land use, and ownership. On the Interoceanic Highway, questions about people's futures have been articulated in relation to claims about who belongs to this space; to what kind of relationship people feel they have to places that are defined by a long history of national and international migration; and to the character of a frontier economy that has experienced the repeated appearance and disappearance of spectacular forms of wealth alongside dire poverty, exploitation, and violence.

The integrative promise of our two roads was thus not only oriented to the question of what could be extracted from virgin forest but also concerned the social implications of the specific relationships between people and things that had previously characterized these so-called frontier zones. We found that the social relationships along the roads entailed an engagement with the tension between the generative potential and promise of the frontier, where people could begin new lives or start new enterprises unhindered by the responsibilities and expectations of prior relations, and the surprising demands that these new lives elicited from the people and environments that continually manifested their histories of prior hierarchical interconnection.

As we explained in chapter 1, roads are primarily thought of in terms of their integrative potential. As network infrastructures, they are conceived in terms of their capacity to facilitate flows of people, goods, and wealth. What becomes erased from the network image of the road, however, are the ways in which roads are also infrastructures that bring people and things into relation with one another, with often powerful differentiating effects. On the Iquitos-Nauta road we saw how the project of connectivity had the effect of bringing together army generals, riverine communities, indigenous populations, and colonists. The road both provoked an interaction, and constituted the difference, between these various social groupings. At the same time it also brought into the mix the question of land: what land is for, what its value is, and who might make a claim to it.

On Route 26, the road has provoked a different set of social preoccupations. Here we found Andean relations running deep through the social life of the road, at times emerging as internal divisions between different Andean regions, at other times playing out as a divide between permanent residents and outsiders, or between extractors of wealth and the custodians of land and of tradition. In the case of outsiders, we described how they are not a singular category, but are defined as much by their often destructive sociality as their place of origin. The question of who is an outsider and who is a *permanente* is an open and contingent one that is never permanently resolved. Even legal papers determining membership in a community ultimately fail to settle the question of whether someone legitimately belongs or not, let alone what the implications of such belonging might entail.

Roads, in this respect, do more than connect places and traverse difference. As infrastructures, they bring people and things together in new configurations where the contours of what is at stake in these configurations is a matter of ongoing negotiation. Out of their connective capacity, roads produce an ongoing politics of differentiation.

Outlining both the integrative promise of road construction projects and the way in which these infrastructures have become participants in the creation of a particular space of cultural differentiation, we hope to have unpacked something of the terrain into which the engineers charged with the job of creating infrastructures such as roads are expected to intervene. In section 2 we change the focus, moving to an ethnography of the road construction process. In particular, we explore how the practices of engineering and construction involve their own politics of differentiation, which are enacted in a constant drawing of distinctions between the technical and the social, the universal and the local, the generic and the specific, and the ongoing attempts to deal with the effects of such distinctions.

PART II

Construction Practices, Regulatory Devices

Chapter 3

Figures in the Soil

[Handwritten annotation: What does soil mean to Peruvians + outsider? How is soil affected by these roads]

Road construction in Peru takes place in unstable material environments. The Andean section of the Interoceanic Highway is beset by frequent earth tremors and by regular frosts that disrupt the apparent solidity of the mountainous terrain, leading to landslips and the disintegration of road surfaces. Along the route there are volcanic springs, gaseous flares, and active glacial flows indicating the ongoing influence of deep geological forces. Floods are common too. In 2001 a deluge of water raised the levels of the Arrasa River, which runs alongside the Interoceanic Highway, to such an extent that a devastating "avalanche" washed away large sections of the old road as well as several roadside settlements.[1]

Our discussions with both local residents and construction engineers often turned to questions of what it takes to settle or manage these dynamic energies. It is well documented that people in the Andes have a deep respect for the landscape as an active and sentient force that is central to the realization of human productive and reproductive potential.[2] But the earth and the mountains with which much Andean anthropology is concerned were not the only spiritual beings to contend with in the space of the road. In our conversations with people about road travel we were frequently told of the dangers of assault by ghosts on lonely stretches of road. Our journeys down the Interoceanic Highway with Carlos and Tomás were punctuated by their memories of things that had happened to them and to others they knew. At the highest pass on the Interoceanic Highway, at Hualla Hualla, Carlos explained to us how important it was to be careful, to be alert to the strange and vigilant against trickery. Someone he knew had met a ghost here. Its deafening screams had echoed all around the hillsides. Terrified, the man had taken refuge at a nearby homestead. The people who lived there were apparently used to such encounters and were quite blasé, telling him that it was "just the devil." They were more worried about the harmful effects of rustlers who came to steal their livestock. Lest we should dismiss the danger and uncertainty of ghostly forces, however, Carlos then told us his own experience of a haunting on that same piece of road. It had been late one night, about 11.30 p.m., and in the distance he had seen the lights of three trucks making their way toward him. Realizing that there would not be room·farther down to pass, he pulled over and waited for them to arrive. Eventually the first vehicle drove by, then the second, but the third never arrived. Carlos said that he had been very tired and may just have fallen asleep. But he was sure that he had not missed the third truck, not least because by morning his young assistant had become very sick, clear evidence that he had been touched by some malign spirit presence.

There is a shrine at Hualla Hualla to protect travelers from such dangers. The breeze-block building is bleak and windswept. At an altitude of fifty-two hundred meters the mountainside is a barren scree of gray rocks and blue lichen, and the clouds hang low over the mountaintops. Water collects in large puddles and reflects the sky, mimicking the glacial flows that nestle in the mountain gullies bearing down on the road from above.

Where the road crests over the top of the mountain pass, rocks have been piled up into leaning cairns by passing travelers. Two green metal tubes have been fashioned into a tall cross, and the face of Christ looks out on the road and down the valley. Most drivers stop here to make a brief offering to the Christ figure, and to the mountain. That day, Tomás and Carlos seemed energized by the thrill of the pass, and as we slowly wound our way down the other side they regaled us with their full repertoire of stories of strange happenings: the strange sicknesses that they and others have suffered, the sudden nauseas, sometimes even deaths. They showed us a lake that changes color from red to green to leaden gray, and places where they have seen plumes of colored flames. These features also relate to hidden earth forces and suggest the possibilities of buried treasure. It is not difficult to imagine the presence of threatening spirits and ghosts at this high altitude, where the air is thin, the volcanic forces are active, and the environment is stark and overwhelming. And yet these ghostly presences are not confined to such spaces. Ghosts and spirits are just as likely to be found in familiar domestic environments.

On one occasion Penny had met up with Clara, an old friend from Ocongate. Clara had spent the day recounting stories about the ghosts that had visited the family home over the years. This house was at the time being run as a small hotel to accommodate some of the senior engineers in the months prior to the start of full-scale construction, when large temporary camps took on this function. It was a time of quite some excitement for the proprietors of these hostels, who suddenly found themselves looking after people who spoke little Spanish—the engineers at this level were predominantly Portuguese-speaking Brazilians—and knew even less about everyday life in a small Andean town. Penny had known this family for twenty years, but she had never been told these stories before. There was something about the presence of the engineers that inspired Clara to remember and rehearse each incident: the time when her brother had felt someone reading over his shoulder, or the even more dramatic time when the sacred robes, which they stored for when it was their turn to clothe the Christ figure in the local church, had been seen floating in the air as if worn by an invisible being. Clara herself had heard voices in the patio talking in a language she did not understand. When she had gone to investigate, there was nobody there. Her father was concerned, and he had suggested that they look for buried bones. Sure enough, when the ground was dug up they found many bones at exactly the spot where the voices had been heard.

That evening, when the engineers came back, Clara was keen to strike up a conversation. She asked one of them whether he realized there were ghostly forces in the house, and whether he believed in ghosts. He laughed, and replied, "No, I believe in math." The brief exchange stayed with us for the way in which it rehearsed an almost stereotypical expression of the different ways in which civil engineers and local people apprehend and engage environmental forces. What intrigued us was the opportunity that this encounter offered for thinking about the diverse epistemological and analytical resources brought to bear on the question of how to understand and intervene in relations with and between dynamic material forces. In considering this issue we are aware of the risk of establishing or entrenching an opposition between a local belief in ghosts and spirits and an engineering belief in mathematics. However, our aim is not to describe such an opposition, but rather to ask how an equivalent analytic interest in material relations ends up on the one hand with ghosts and on the other hand with mathematics. An assumption of equivalence allows us, we suggest, to interrogate the ways in which a technical understanding of material relations operates, like ghost stories, as a specific means of engaging and intervening in relations with unruly and uncertain material worlds.

Although anthropology has perhaps been more often associated with the study of the spiritual, religious, and ghostly as domains of belief wherein "culture" might be found, there has also been a long-running interest in anthropology in the cultural dimensions of scientific and mathematical knowledge and the ways in which numbers in particular are mobilized as a foundation for an epistemological engagement with the world. In recent years, several prominent scholars have articulated strong critiques of universalist conceptions of mathematics and drawn attention to the huge variety of ways in which numbers are used to represent and to effect particular worlds.[3] This work on the anthropology of number suggests that to understand numbers' social importance we should approach numerical entities not as universals, but as specific relational forms. In these approaches a contrast is drawn between approaches to numbers as passive forms, used to construct objective accounts of external realities, and approaches that are interested in what numbers do, in the relationships they entail and the worlds they create. Scholars such as Verran (2001) and Lave (1988) suggest that in order to understand the power that numbers play as a resource for social and cultural differentiation we should pay greater attention to the practices through which numbers are generated and through which they acquire social and political force. That is, we need to know what numbers are made of and what they in turn make happen.

In what follows, then, we turn our attention to the performativity of numbers in road construction projects. Following Andrew Abbott's suggestion that the specific ways abstract knowledge is deployed mark the jurisdiction of particular professions (Abbott 1988), we begin our ethnographic exploration of civil engineering practice with engineers' interest in numbers and their belief in math. Civil engineers clearly distinguish themselves as an occupational group from many of the other kinds of workers engaged in road construction. At the same time, there are many branches of civil engineering, and in the following chapters different specialties will come into view at different times. In all cases, though, we can safely assume that a commitment to mathematics, and an ability to mobilize numbers to produce particular forms of abstract knowledge, distinguishes the engineers from those who orient themselves to ghosts and spirits. However, we were also delighted by the engineer's affirmation of a "belief" in math. For here we are given a concrete point of departure that resonates with our ethnographic experience; that is, a sense of trust and investment in what it is that numbers can do. Attentive to the ways in which civil engineers tack between performative and representational idioms, between realities effected and realities represented (as Muniesa [2014] so neatly puts it),[4] we set out to engage the technical dimensions of infrastructural transformation and the centrality of pragmatic abstraction in engineering knowledge practices.

Planning

On the Interoceanic Highway the process of building the road had begun long before there was any sign of movement of earth or labor along the highways in the departments of Cusco or Madre de Dios. It is impossible to locate an originary moment for the road construction project, but a significant early moment in the process was April 2004, when plans for the highway were ratified in the Peruvian parliament.[5] The ratification was made on the basis of the findings of a prefeasibility analysis (*estudio de prefactibilidad*), which had been conducted by the Ministry of Transport. The study had outlined the current state of the roads in southern Peru, the economic benefits that would accrue from improvements in the road network, and the costs of such a project based on standard figures drawn from the World Bank. This needs analysis concluded that a road construction program in southern Peru was technically, economically, and environmentally viable. It made various recommendations as to the form the road construction project should take, including (1) that construction should proceed in three discrete and concurrently running projects that would focus on different sectors of the highway, and that each of these subprojects should be conducted in a number of predefined stages to ensure the road remained useable during construction; (2) that innovative forms of public–private financing should be pursued; and (3) that the development of the road should be aligned with environmental and social development aims in the region.[6] On April 29, 2004, the Peruvian parliament approved this document, allowing for the initiation of what was generally referred to by our interlocutors as the first stage of the road construction process: the carrying out of a full feasibility study (*estudio de factibilidad*) for the project.[7]

In the offices of the Ministry of Transport and Communications in Lima a team of bureaucrats and scientists employed by the ministry and a number of engineering companies, known as the Consorcio Vial Sur, had begun the work. The study was the first to be conducted under a new arrangement of public-private partnerships to fund infrastructural projects in Peru,[8] and its purpose was to present robust multidisciplinary evidence to demonstrate both that the road was necessary and that the proposed routes were the most appropriate. In this respect the study was an exercise in transparency with respect to distributed decision making, but it also necessarily constituted an engagement with the same unstable and mutable landscape with which we opened this chapter.

The seven-volume study that resulted was an impressive collation of numerical data, maps, tables, charts, photographs, descriptions, and analyses of the road in its various past and possible future manifestations. It brought together existing data on the state of the road network in this region of Peru with new and specific data gathered to help indicate the best route for the new road. This choice of the best route had to take into account the relative needs, costs, and practicalities associated with each of the three options suggested in the prefeasibility study. It involved the collection of data on transport plans, geometric design and signaling, the geology and material makeup of the land, the local availability of natural resources for use in road construction, hydrological systems, possible road designs, and the basic social and economic profiles and trends of the region.

As might be expected, the study that emerged was the outcome of an engagement with a wide array of different standards and regulations, which dictated how data could be produced and what these data would mean in relation to comparable data from other places and times. We suggest that, in this case, the power of the numbers in the report to maintain their legitimacy in the eyes of experts and publics alike rested on the relationship between the specificity of the data presented and the generic power of the standards in relation to which they had been produced.

As a public institution, the Ministry of Transport was subject to a series of legal regulations dictating the terms within which road construction projects could be pursued. These ranged from the most general labor and employment laws aimed at mitigating potential corruption or ethical misconduct to specific laws governing road use and construction They dictated, for example, the weight and number of passengers vehicles may carry when traveling internationally, the provision of ducts and chambers under any new road where future telecommunications wires can be laid, and systems of standardization, such as manuals dictating the load-bearing capacities of different categories of roads. It was on the basis of these normative regulations, and the standards that they supported, that claims about the potential profitability of the road, its social impact, and its usefulness could be made. For example, projections about poverty were determined through the use of data collected by the Peruvian Ministry of Economy and Finance that were published in the *Marco Macroeconomico Multianual 2005–07*, which illustrated macroeconomic projections for the years during the initial phase of construction. Data about the potential profitability of the road was interpreted using two frameworks: the World Bank's *Highway Design and Maintenance Model* (HDM-III-1995), which provided data on the projected deterioration of the road under different levels of usage, and the *Metodología del Excedente del Productor*, which could determine the potential usage that was likely to result. The combination of these measures allowed the study to produce figures that projected benefits such as an annual increase in tourism worth $15 million a year and a reduction in interruptions in transit worth $13 million a year, thus demonstrating the project's financial viability.

Regulatory enforcement of the use of standards thus played a central role in orienting the look and feel of the feasibility document that was produced and providing legitimacy to the numbers that appeared there. Civil engineering itself, as a pragmatic science, relies heavily on an understanding of systems of standardization, from the basic measurement of length, weight, and force, to more complex standards that are built in to the software programs engineers use to make predictions and forecasts, which inform the direction of their construction. Standards can be seen as an important way in which engineering practice becomes entangled with a technopolitics of control that works to determine proper and appropriate conduct in the production of data,[9] transforming our understanding of engineering from a merely technical exercise into a political one.

For the engineers we got to know, the power of the numbers they were working with came from their capacity to mediate between the specific social and material circumstances found on the ground and the legal, financial, and engineering standards to which infrastructural projects had to adhere. That standards are central to engineering practices was indisputable; several of the engineers we spoke to put the mastery of standards of different kinds at the center of their professional expertise. To be an engineer is to be able to understand and navigate the proliferating systems of standardization that enable calculations and predictions to be made and, ultimately, things to be built. Standards allow for figures to be evaluated for their accuracy, workability, and usefulness, and in this sense they have a determining relationship to the look, feel, and functionality of the built environment. Moreover, in their capacity to produce a normative effect, standards are also tools that allow for the production of legitimacy.[10] The politics of numerical production in the feasibility documents was thus intimately linked to the production and adherence to standards. These standards enabled those writing reports to produce aggregated figures about the constantly shifting and mutable social and material environments where they intended to make an intervention—figures that could be used as the basis for projections about the future viability of the road. In the case of the feasibility study, then, it was only within the terms of the standards that were openly acknowledged as relevant and pertinent that the producers of the study could claim that their figures were "correct" and "accurate," that their projections were appropriate, and that the study was able to provide the basis for decision making as to how to proceed with the road-building program.

Although the production of the document was an important act of political as well as technical alignment, in hindsight the numerical data it contained ended up being surprisingly fleeting. The feasibility study alone was reported to cost twelve million Peruvian soles (US$4 million), an indication of the time and effort invested in this document. However, once the study was finally completed and the contracts awarded for the road construction on the basis of the feasibility study, the document seemed to disappear from view. By the time the road construction process started in earnest, we discovered that this huge tome had already become a historical artifact, gathering dust on the bookshelves of offices, stored on the hard disk of computers in the Ministry of Transport, and filed in the archives of town halls along the proposed route. Just two years after the study's completion, the engineers employed by the road construction program told us that no comprehensive engineering study of the highway had ever in fact been completed. Indeed, at the point when we begun our ethnographic work on the road construction process, the engineers working on the Interoceanic Highway project were starting from the beginning, once again producing detailed mathematical data, this time to orient the construction process itself.

Far from being a failure of the feasibility study, the amnesia it seemed to produce after the event of its production is indicative of the role such analytical exercises play in the course of large-scale public infrastructure projects. As we have already noted, the purpose of the numbers contained within the document was not to demonstrate a universal and stable truth; rather, it was to provide a link between the specific contexts of this particular road construction project and the projections on which the social and political viability of the project could be established. Anna Tsing (1999) has suggested that such is the power of numbers to produce political effects at times it is not even necessary for there to be substance or truth behind the numbers. She calls this capacity of numbers to mobilize people and materials a "conjuring" effect. Like the appearing and disappearing rabbit in the magician's hat, the reality that numbers conjure up is tangible and yet elusive and of uncertain provenance. To make something visible does not make the conditions of its emergence transparent. In spite of the agreed accuracy of the figures produced for the feasibility study, and the acknowledgement that the standards used to give the study legitimacy were appropriate and correct at the time it was conducted, it was soon apparent that the analysis that preceded these figures could not be used as the basis for the actual construction of the road. This suggests that the figures produced in the feasibility document were primarily producing the road as an idea. To produce the road as a material form required quite different kinds of information and the deployment of different modes of abstraction. In both the prefeasibility and the feasibility studies numerical data and mathematical formulae played a key performative role in enabling the state to enter into contractual agreements with banks and construction companies in the name of their citizens, and in open adherence to legally defined standards.

The political force of the figures produced through this work of analysis and description should not be underestimated. In democratic polities, public infrastructures need proven public utility. The public good has to be demonstrable, and the "public" itself also has to be conjured up as a material presence alongside, and in relation to, the figures. Publicity is thus a further element that surrounds construction projects. Regular updates were given in the local papers during 2004 as to the progress of the data collection—in June, 58 percent complete; in August, 65 percent complete; by September, 70 percent of the data collection and analysis had reportedly been carried out. The construction consortium also held public workshops in the towns along the potential routes, to "consult" and to "inform" people about the value of the new road, its benefits to them, and the role they would be expected to play in bringing about a new and improved future.

These meetings and the many PowerPoint slides, maps, tables, and figures they contained did not prevent anxieties and questions from dominating these consultation exercises. People were concerned from the very beginning over how these projected figures were to be achieved, and by the suggestion that somehow it was their duty, as individuals and communities, to reimagine themselves as players in international markets who were ready to exploit the new business opportunities that the road would somehow provide. Nonetheless, the figures did set the terms of this engagement and were key to the ways in which the road project unfolded, in ways that we trace in subsequent chapters. In this respect, the figures had the effect of conjuring the road into being as a particular kind of idea, one that took its form from the numbers and standards through which it had been configured in the studies and the consultation exercises.

One way of interpreting this performative enactment of the figures is to consider them in the vein of what Annelise Riles, following Vaihinger and Ogden (1924), has called a subjunctive position, or a philosophy of "as if" (Riles 2011). Describing the activities of swap traders in global financial markets, Riles has shown how the knowledge practices of financial analysts are oriented less to the task of capturing or stabilizing a knowledge of how markets are—an engagement with the material properties of things—but rather work by mobilizing algorithms that assume how the market might operate, which allows traders to make immediate decisions on that basis. In a similar way, the standards we have discussed here had the effect of allowing the numbers in the report to be stabilized long enough for predictions to be made as to what the effects of building a road would be on the local environment and on economic conditions if the numbers were to be utilized in the road construction process itself. We might say that these standards and regulations operated as "useful fictions,"[11] allowing for the appearance of a context for action, while retaining the knowledge that the report's figures would never be the basis upon which the road would actually be built, because by the time the road was built, all kinds of things would have changed. The artificiality of the context that allowed for predictions to be made was the result of having to project a complex and multifaceted scenario into an ultimately unknowable future. Nonetheless, the systems of standardization allowed these projections to be made by inhabiting the subjunctive position, operating "as if" the standards being adhered to were actually fully aligned with the world within which they are produced.

At the same time, this subjunctive philosophy of "as if" appeared to open up the potential for suspicion and ambivalence over whether the conditions would hold long enough for the project to actually materialize. For Riles's traders, these conditions were constantly deferred into a future so that the effects of holding off the potential of closed knowing could produce a space within which financial effects could accrue and profits could be made. The road builders, however, were patently aware that the day would come when the philosophy of "as if" would necessarily be stalled by its materialization into a physical road with discernable, if not ultimately describable, effects. As we have seen, many ethnographies of expert practice and the use of numbers have become enchanted with how numbers work as forms of abstraction, carrying and producing meaning in dematerialized ways, with powerful political effects. However, focusing on numbers as abstractions risks erasing the material practices that in many respects work to stall the potential of numbers to become abstract worlds unto themselves. In what follows we show how in engineering the philosophy of "as if" works as an explanation for the power of numbers to tame and manage unstable material worlds only if we specify the feasibility study as a particular phase of construction within which such numerical practices apply. However, although the feasibility studies were perfectly adequate for the job they were produced for, they could not form the basis of knowledge of how to actually build a road once the financing was released. Once the road construction phase began in earnest, a wholly different relationship emerged that reconfigured our understanding of the work that mathematics does for engineering, from a philosophy of "as if" to what we call a philosophy of "as long as."

Material Preparations

In August 2005, CONIRSA, a consortium of one Brazilian and three Peruvian engineering companies, was awarded the contract for the improvement and widening of the highway between Iñapari on the border with Brazil and Urcos in the region of Cusco. This would complete the paved road connection between Brazil and three ports of San Juan, Matarani, and Ilo on the Peruvian coast, a program of work that became known as the Interoceanic Highway project (see map 2). It was also in August 2005 that we began our research on this road. Whereas the first few months of the contract were characterized by a perceived lack of activity on the ground (as the various studies were being completed and compiled), by January 2006 CONIRSA had a visible presence in people's lives. The first thing the consortium had done was to take over responsibility for maintenance of the current road from the Ministry of Transport and Communications. Heavy machinery emblazoned with the logos of the contracted firms was moving up and down the road, spraying water to manage the dust thrown up by the trucks and buses during the dry season and repairing the worst parts of the highway damaged by the previous year's rains. Meanwhile, CONIRSA had also begun a phase of design engineering that would allow them to move from the feasibility study phase to the road construction phase. The new site office in Cusco, a converted residential house crammed full of papers and computers, was a hive of activity, with engineers and their drivers bustling in and out of the building, their conversations interrupted by ringing phones and bleeping mobiles. Along with their main headquarters in the city, CONIRSA had also set up a series of field laboratories where detailed analyses of the state of the road and the material surroundings were being carried out to determine how construction should proceed.

The CONIRSA laboratories were our first encounter with the practical work by which civil engineering uses mathematics to construct a material profile of infrastructures that are under construction, and the relationship between these infrastructures and the environments into which they are being placed. In the labs we heard the engineers distancing themselves from the plans of Ministry of Transport and Communications bureaucrats, contrasting the ambivalence they felt toward these preliminary government studies with the material immediacy and powerful tangibility of their own analytical work.

Field laboratories are the sites where the detailed analytical work of road construction takes place. They are also the places where laboratory engineers learn their trade. Few universities in Peru have engineering laboratories, and so, as one engineer put it, 95 percent of the practical life of the laboratory engineer is learned on-site, in field laboratories like those set up by CONIRSA. Although standard mathematical techniques can be learned in the classroom, it was clear that the labs were the place where the real work of learning to be a construction engineer took place. What was it, then, about the practices found in the labs that differentiated them from the generalized abstractions that could be taught in the classroom? What kinds of mathematical operations were taking place in the field laboratories that were unavailable in the hypothetical environment of the university, and what might this tell us about the relationship between the "as if" and the "as long as" approaches and the specific modes of abstraction through which construction engineers establish their expertise?

On the Interoceanic Highway, the first field laboratory was set up in an empty house on the edge of the town of Urcos, conveniently located directly across the road from where the section of the Interoceanic Highway from Urcos to the Brazilian border would begin. Armed with standardized measurement criteria in the form of tables, maps, and measuring instruments, teams of technicians were sent out up this section of road to gather raw materials for analysis. Every two hundred meters, a section was scooped out of the road, and the materials were brought back to the laboratory, where these samples were transformed and manipulated to produce a picture of the physical makeup of the highway as it currently existed.

We watched as the bags of mud were brought into the laboratory and a couple of lab technicians wandered over to attend to them. First, the samples were emptied onto the ground and left to lie in the sun to dry out. Once the samples were dry, the workers began separating out the clumpy mix of rock, sand, and dust into gradations of different sizes. Kneeling on the floor the technicians began by attempting to pass the largest rocks through a metal

Figure 6. Sorted materials in the laboratory

grid with large holes. Turning the stones this way and that they worked to discover the geometrical limits of these objects, until they were satisfied that they were unable to fit through the preassigned holes. Those that did not fit were put to one side, and the rest were moved on to the next stage. Then a grid with smaller holes was used to separate out the next largest stones from the rest of the sample. Once again, the technicians tried to push the stones through the holes, teasing and easing them but being sure not to break them in the process. Once the hand-sized rocks had been sorted, the rest of the sample was shoveled into a large sieve, which was shaken with a rhythmical tapping and swirling action for a set length of time. At this point, the materials left behind in the sieve were again poured out into a pile on the ground, and the process was repeated with a slightly smaller sieve. Over and over, the sample was swirled and shaken, creating a haze of dust in the light wind, until a series of different sized piles gradually appeared, laid neatly out on the floor. Finally, after the smallest sieve had been used, the remaining dust was swept up and brushed away.

Figure 7. Testing the swelling properties of soil samples in the laboratory

What we were observing here was the first stage in a longer process of the classification of matter for particular ends. This first stage entailed a separating out of a continuity of mud and stone into discrete gradations based on relative size of particles. Once the mud and rock was separated out into these piles it was bagged up. Standing back and looking at the bags of material, the two technicians set about putting each of the bags in size order, with the largest bag on the right and the smallest on the left, shuffling them around until they were neatly lined up in order of relative mass. Through sensory engagement the matter that had been separated according to a standard measure was ordered into a scale of greater and lesser quantities.

Engagements with Qualities

If the process of classification these engineers were engaged in started off with an ordering of matter, it did so with the broader aim of determining not only abstract and continuous details about the soil—its weight, volume, and particle size—but also the dynamic qualities the earth would be likely to manifest. One of the most powerful effects of the material engagements that engineers encounter in the field, as opposed to in the classroom, is an ongoing relationship with material qualities and their pressing social effects. When engineers working on the road talked to us about the materials they encountered in their work, it was precisely the qualities of the materials that provided the basis for their engagement. We often heard engineers talking about the battles they had with the earth—mud that was like an undulating mattress; capricious water spouts that, however hard you tried to tap them, would reappear in different places; recalcitrant webbing that would not unfold; and concrete that would not set. The specificity and challenge of these material qualities were what provided the need for field labs. The job of the laboratories was to transform this site-specific and embodied experiential knowledge of substance in action into a mathematical description that would make the terms of its unstable qualities knowable, and the possibilities for its stabilization calculable. First of all, as we have seen, the samples underwent a process of disaggregation or dispersal into parts determined by particle size, which, when reaggregated, told the engineer about the particular composition of the specific sample found in the field. Once the samples of earth had been classified according to their specific constitution, further tests were carried out on these subsamples to determine how the materials would react to the weight of vehicles under different weather conditions.

One such test that we watched being carried out in the field laboratory was the plasticity test. The aim of this test was to determine how supple the soil is under specific controlled conditions. Unlike the first test, which took place in the open air under the flimsy cover of a temporary awning, this test required a series of enclosures—the laboratory itself with its four concrete walls, ovens in which samples could be dried, and humidity chambers in which they could be left. Bruno Latour (1999) has characterized the construction of field laboratories and the translations required of the substances in order that they are understandable within the laboratories as a method of "gridding." For Latour, processes of enclosure and inscription "grid" materials into particular relations with one another, allowing them to be stabilized in such a way that they are then able to travel. However, in these engineering laboratories we were struck less by the stabilization of matter into a gridded form, although this was undoubtedly a part of the process, and more by the powerful material aesthetic that emerged in the course of the process of analysis. In the act of producing numbers that could be inscribed in notebooks and transferred onto computers, the engineers and technicians found themselves making new relations with materials as much as they ended up abstracting them.

The room where the plasticity test took place was quieter than the sorting space, and the technicians stood at workbenches with their backs turned to one another, their bodies focused on the detailed work of analysis. The technician showing us the plasticity test demonstrated first how the soil was prepared by being sifted and ground into a powder of particles of a predetermined size. The sample he was using had already been prepared by being dried in an oven for a set length of time. The technician weighed out a prespecified amount of the dry soil. Once it was weighed he took a measuring flask and poured a small amount of water onto it. All of this took place at a much smaller scale than the separations we saw earlier. In contrast to what had been a separation out of kilograms of material, the plasticity test was being conducted on grams of matter.

Once the earth and water was mixed together, it was taken in a little container over to a device that we were told is used to determine the point at which enough water has been added to the sample. This is the liquid limit (*el limite liquido*).[12] Very carefully, the technician scooped out a spoonful of the wetted soil and placed it in a small copper dish mounted on a wooden stand. He then smoothed out the top of the sample until it made a neat penny-sized disc in the copper dish. Taking a thin rod between his fingers, he leaned forward and slowly marked an indentation down the middle of the sample, separating the two halves by a gap of a few millimetres. Holding the wooden base the copper dish was resting on, he then used his other hand to turn a small handle on the side of the machine. Through a system of cogs the copper dish was raised up a couple of centimeters above the wooden block before it suddenly dropped back down upon the block with an abrupt click. The force of the drop was such that it was immediately clear that the two sides of the sample had moved slightly toward one another. This was not a good sign. The aim was to get the mixture to a consistency that the two sides would only meet after thirty to forty drops—too few, and this would mean there was too much water in the sample, and the earth would have to be redried and the process begun again; too many, and more water would need to be added. In this case, the technician's suspicions that the sample was too

Figure 8. Preprepared soil samples in the laboratory

wet were right—five or six drops later and the two sides of the coin-shaped sample had touched. The test was over for now, and would have to be repeated with a new soil and water mix. The technician explained that this experimentation was part of the process of discovering the qualities of the soil that he was working with. He explained that once they have been able to determine, through repeated testing, when the soil sample achieves the right *limite liquido*, the amount of water that has been added to achieve this level is written into the notebook, thus stabilizing this piece of knowledge about the mud, and allowing the second stage of the test to begin.

However, the process of testing does not take place in a relational void. As pointed out earlier, an awareness of the qualities of the materials under analysis precedes their appearance in the lab. Road projects are characterized by engineers in terms of the material challenges they throw up—earth that is too muddy, a lack of sand, an abundance of rock, the presence of earthquakes, and so forth. The tests that take place in the laboratories engage prior expectations of the qualities of the materials undergoing the test. Encouraging the materials to manifest their particular qualities in numerical form is as much a process of coaxing the numbers out of the matter as it is a process of subjecting materials to a series of standardized tests, which abstracts materials from their contexts. Mathematics was thus, in this case, not the means of revealing or determining wholly unknown qualities of soil; it operated more as a method of rearticulating the properties of matter that engineers were already engaging with on a day-to-day basis and that were already knowable in other ways.[13] As Latour (1987) and Knorr-Cetina (1999) have observed, scientific practice involves acts of inscription and writing to stabilize the worlds being constructed in spaces like laboratories. This is an important means by which substances are extracted from their life histories and other affective qualities. However, in the road laboratory, the process of inscription, although present in the act of analysis, seemed to be an aside to a face-to-face encounter between men, materials, and machines. The placement of the notebooks being used in these field laboratories was telling: outside, the notebook was set on the edge of a wall, the technician having to turn round 180 degrees to scribble a number in a column on a flapping page before turning back to the mud and rocks themselves. In fact, so marginal was the notebook to the activity in hand that at one point it got lost, and everyone was mobilized to scramble around and find it. Inside the labs, the inscriptions again took place at arm's length, to the side and away from the encounter between the technician, the device, and the substance being measured. Moreover, computers onto which the notes were then transferred were in an office in another part of the building entirely.

As if to emphasize the importance of material qualities in the process of analysis, Hannah was taken over to another bench where someone else was doing the second stage of the plasticity test, which was to work out the plastic limit (*el limite plástico*)[14] of the soil:

> A man sat rolling balls of what looked like plasticine into little snakes, something I had seen previously in other road labs and that always looked so incongruous in relation to the grand gestures of inauguration, planning and financing that is the public image of road construction. I was soon disavowed of my confusion as the technician began to explain to me the chain through which rolling plasticine could be re-imagined as an important part of road construction. To work out the *limite plástico* of any soil, the method that they use involves rolling the soil/water mix under the tips of the fingers on a glass

plate, into 3mm wide batons. The technician told me that he was able to estimate from experience what a 3mm width looks like as he rolled the soil sample out and didn't need any kind of measuring device to determine the width was correct.[15] Once the batons have been made, they are placed in a container to dry out and gradually cracks begin to appear. Very sandy samples will just disintegrate upon rolling, as happened when I was being shown the test. The technician gave me a bit of the sample to rub between my fingers and to feel how grainy it was, sharing with me the tactile experience of a sample with no *limite plástico*. For those more clay-based soils that do hold together, I was told how they are re-weighed at the moment where cracks appear and then dried out again and weighed to work out the water content at the *limite plástico*. To determine the final figure for plasticity (*plasticidad*), the percentage of water at the *limite plástico* is subtracted from the percentage of water at the *limite líquido* (liquid limit) to produce a percentage measure of plasticity. The technician then showed me some results from a test that had already been completed. In this test the material had been 30 percent water for the *limite líquido* and 22 percent water at the *limite plástico*. This made the *plasticidad* of the material 8 percent. (Extract from Hannah Knox's fieldnotes, 15th March 2006)

In this test the soil was being put into interaction with another substance— water—in order to understand the relational properties it manifested under particular conditions. Unlike the way in which the numbers in the Ministry of Transport documents were oriented toward the politics of funding a new road project, the numbers that we found in the laboratories were oriented to the quite different aim of making a physical structure that would withstand certain effects. The math required to generate this effect was not the hypothetical algorithms of Riles's traders, but it was still importantly provisional. Unlike the scientific laboratories described by Latour, where the aim of description and inscription was oriented toward a search for robust truths, what we observed in the field laboratories of the road construction process was a practice oriented toward the specific pragmatic end of making a material structure that could endure within the parameters of particular conditions. The mathematics that was mobilized was likewise not absolute but was oriented to this contingent end.

In his semiotic analysis of mathematical process, Brian Rotman (2000) identifies the way in which mathematical actions rely on a conception of the mathematician in three different modes—the person, the subject, and the agent. Put simply, for Rotman the person is the embodied and relational being with a life history and emotions and experiences who, in order to do math, imagines him or herself into the role of the subject who carries out mathematical procedures. The mathematical subject has the technical capacity to think up and imagine mathematical formulas and inscribe them on paper using notational devices that posit objects and relations between them. The mathematical subject is also the imagined interlocutor who receives instructions, such as "add," "consider," and "let x be the case," in the syntax of mathematical language. However, due to the nature of mathematical formulations it is often impossible for the individual mathematical subject to prove the validity of a mathematical theorem, as the circumstances under which it would need to be tested are multiple if not infinite. In order to allow a proof of sorts to exist, the mathematical subject has to imagine the existence of a hypothetical agent who, under conditions of infinite time or space, or both, would be able to demonstrate the universal validity of the theory.

Rotman's characterization of mathematics in general is useful for thinking about the role that math played in the experiments we observed in the field laboratories. As we pointed out earlier, the mathematical calculations mobilized in the laboratories were not so much an abstract description of a previously unknown world, but rather provided a means of redescribing, or translating, a relationship that was already known in other ways. To put it into Rotman's language, the numbers were being produced and analyzed by the engineer as subject, but in a manner that was intimately related to the engineer as person, already embodied and enmeshed in a relationship with the materials under analysis.

At the same time, the *agent* that Rotman identifies as central to mathematical practice was also present in these practices. The very act of sampling the road for detailed analysis at two hundred meter intervals would seem to provide a discontinuous picture of the road, as if it were only known at these single points. However, the way in which mathematics was made to work, by mobilizing an agency that was beyond the capacity of the calculating subject, allowed the gaps that appeared through a merely subject-based analysis to be overcome. Sticking with a subject-oriented analysis of engineering would lead to a position where for a picture of the road to be determined the whole road would have to be dug up and analyzed—an action that would, of course, in a situation reminiscent of Zeno's paradoxes, entail the irrevocable transformation of just that object that was supposed to be undergoing analysis.

Instead, thanks to the capacity of the mathematical agent to fill in the gaps and allow a smooth, generalized picture to appear via numerical abstraction, most of the road could be left untouched while a working knowledge of what it was made of could nevertheless appear.

The mathematical agent thus allows a provisionality to enter the calculations. In the case of the planning documents this provisionality took the form of the subjunctive relationship to the future, which we described earlier, allowing standards to hold together a road construction process that was seeking political ratification. Likewise, one of the things that we learned from the engineers we were working with was that they were not in pursuit of a total knowledge of the road in the way in which an analysis of engineer-as-subject might suggest. Yet neither were they interested in relating to numbers as mere abstractions that would allow them to make hypothetical calculations for political effect. Rather, they were interested in producing an understanding of the road that would allow them to build a road that would fulfil the obligations they had signed up to in their contractual agreement with the state. Although an ideal road remained a hypothetical possibility, all were aware that an ideal road was not what they were working toward. Instead, they had been charged with the responsibility of building a specific road that would not corrode so quickly that it could not be maintained; that would not exceed the budget in such a way that the project itself would risk being left uncompleted; that would improve the livelihoods of people living alongside the road, but only insofar as they also needed to find ways of improving their own standards of living; and that would mitigate environmental damage, but only as far as that environmental damage could be shown to have been caused as a direct consequence of some otherwise avoidable activity. It is in this sense that we suggest that engineering practice proceeds less on the philosophy of "as if," although this is a dimension of such practice, and more on the basis of what we are calling a philosophy of "as long as." Again, we see how the mathematical knowledge production through which road building proceeds is successful "as long as" it leads to the fulfilment of certain conditions the engineers are striving for.

Importantly, the analysis the lab technicians were conducting was oriented not just to the production of knowledge about the road as an end in itself. Rather, it was expected that the analysis we observed would produce knowledge that could be mobilized to the end of generating new materials that could be used in the road construction process. Although a large part of the process of analysis in road building is determining the qualities of the materials that are to hand, equally important is the more experimental work of recombining materials with particular qualities to produce substances with an optimal capacity to relate to the environments in which they will be put to work. If the relevance of the "as long as" mathematics of the lab for producing numbers that have a social and political force is to be understood, we must also consider how engineering is oriented in practice not just to the production of data about existing material qualities. It is also oriented toward this recombinatory exercise of producing new kinds of matter out of the available mud and stone and working out how to put that matter to use in the future. In this respect we show how the philosophy of "as long as" is not simply a facet of mathematical calculation but orients other kinds of relationships that are also necessary for the building of a road.

Making Substances

On the Iquitos-Nauta road, one of the main material challenges that engineers were facing in their attempt to deliver the road was the instability of the mud. Tests had revealed the mud to be highly plastic, meaning it provided a very unstable surface, a fact attested to by the near impossibility of walking around in wet weather because of the way the mud formed huge, unwieldy clumps on one's boots. The challenge was how to find a mixture of substances that could be laid underneath the asphalt to produce a stable base. Penny visited a plant near the road that was built to assemble the specific combinations of materials (sand and stone) that the labs had modeled as providing the optimum base for the road.

Initially the best possible proportions of mud and sand were calculated on computers that modeled the effects of putting different materials together. The data entered into these models derived from the analytical work that we saw conducted above, which in these calculating machines became subject to algorithms that linked together the properties of different kinds of matter, producing visual depictions in graphs and charts of the relative effects of different combinations. As with the laboratory tests, the engineers' relationship with the computer programs was one of tweaking and adjusting until, as they put it, the visual and mathematical output "felt just right." These combinations produced in data were then used as the basis for further laboratory tests where numerical predictions would be able to be rematerialized into tangible substances that could be subjected to experimental methods.

Once a particular mixture of materials had been determined, first through this combination of numbers and then of substances at the miniature proportions of the laboratory tests, the proportions were scaled up for the construction process. Large machines mixed together preprogrammed quantities and sizes of stone and sand in appropriate proportions for use in the different layers of the road. Construction engineers explained to us that, as far as an engineer is concerned, the road has three layers—the cap, the base, and the sub-base. Each layer of the road required different predetermined properties. In the case of the Interoceanic Highway, the engineers were working to U.S. standards outlined by the American Association of State Highway and Transportation Officials—a standard notable for requiring that the weight of vehicles that would travel on the road be calculated as weight per axle rather than weight per vehicle. It was through the explicit use of such standards that engineers were able to ensure that all national roads could be classified according to a correspondence between their constitution and their projected usage.

From the engineers' perspective, the quantitative combination of mud, stone, and sand was something they felt was more or less under their control. The combination of what Rotman identifies as person, subject, and agent allowed the calculations to have a legitimacy and force, providing a robust basis for belief in the capacity of the numbers to orient decisions about road construction. However, stories of the challenges of road building were replete with accounts of other considerations that did not fit the triple-faced identity of the mathematically oriented engineer, and which were therefore described as impinging from the outside on the business of engineering proper.

One of the problems we frequently heard about was the practical difficulty of getting ahold of the materials that the numbers dictated were needed. For example, on the Iquitos-Nauta road an ongoing challenge facing the road builders was a lack of sand in the region suitable for road construction processes. Ironically perhaps, the fragile ecosystem alongside some stretches of the road was particularly in need of protection because the soil that sustained the forest plants was uniquely sandy. Some said that this was due to the fact that about fifteen million years ago the Caribbean Sea stretched down from the north to this point in the Amazon. However, this fine, white sand, which sustained these local forests, was neither suitable nor available for road construction processes. To find the sand needed, the engineers had to search up and down the river for potential suppliers. In the end the suppliers they came across were located two weeks upriver from the construction site, an issue that caused the engineers a great deal of consternation as they fretted over the extra burden that transportation of raw materials down the river on barges would put on the finances and time frame of the project.

Likewise, on the Interoceanic Highway the question of how to source materials opened up the detailed numerical calculations that we observed in the labs to a whole series of other considerations that had to be incorporated into decisions about how the road would be built. In the highlands, the land was solid and the mountains provided the stone from which different gradations of materials could be manufactured using machines known as *chankadoras* that crushed large rocks into gravel and sand. On the long stretch of the road in the lowlands, however, a similar problem to that faced by the Iquitos-Nauta road was encountered. Whereas here there was plenty of sand, this meant the land was unstable and prone to flooding. One of the problems faced by previous travelers along the road was the unpredictability of the route during the wet season, when the dirt road would turn into a fast-flowing river. At certain points on the lowland stretch, a sudden downpour could transform a small fordable stream into an impassable torrent. To ensure that the new highway would be able to mitigate these problems, the engineers realized that they were going to need large amounts of rock to raise the level of the road for stretches of several kilometers. Under the raised road they would have to place drains at regular intervals to allow the water to flow through. Working out how high the road needed to be and precisely how water would be diverted so it did not worsen the problem of flooding for farmers living alongside the route provided the basis of a whole new round of calculations, involving not just lab technicians but topographers and engineers trained in the art of drainage, concrete construction, and bridge building. Moreover, it turned out that achieving these raised highways was also going to involve the participation of the community relations office with a combination of legal and economic expertise and negotiating skills that we explore in more detail in chapter 6.

Even though the mountains of the Andes provided copious amounts of stone for construction teams in the highlands, bringing the necessary materials down the currently tortuous route from where they could be sourced was, as on the Iquitos-Nauta road, seen as uneconomical and time consuming. The search began for alternative supplies of materials along the river system in the southern Peruvian jungle. An unlikely source of stone soon presented itself in the form of the by-product of the gold mining taking place along the region's rivers. As we have already touched on elsewhere, the rivers in this part of the Amazon, flowing as they do from the foothills of the Andes, have long been rich fields for artisanal and semi-industrial gold collection. Environmentalists, dismayed at the havoc caused by unregulated gold mining, pointed to this industry as one of the key reasons why the whole Interoceanic Highway project should be rethought. They feared that opening up a new fast route into the lowlands would lead to an uncontrollable expansion of gold extraction, with irreversible effects. It is somewhat apposite, then, when the need for stone for road construction became apparent, that the road construction team would turn to the gold miners themselves.

Alluvial gold mining operates by scaling up the principle of hand-panning to a semi-industrial level. Whereas the romantic image of the gold prospector has the panner standing up to his knees in a crystalline stream, scooping clear water through a sieve and gathering the twinkling drops of gold that are left behind, the realities of gold mining in the Amazon are much more brutal. Most of the miners in the Madre de Dios region have left their families in the highlands, and traveled down the road on the top of a fuel truck to find work in small-scale, and predominantly informal, unregulated mining enterprises. Mainly from agricultural communities, these miners are used to working with the land. However, whereas agriculture in the Andes involves a long-term and intimate connection with a historicized landscape, mining for gold involves short-term, temporary installations on the edges of shifting rivers. Fortunes change over time as rivers become exhausted and rumors of more profitable sites open up elsewhere. Mining camps are temporary, flimsy affairs, normally just a series of shacks made with wooden poles over which blue plastic sheeting is strung to keep the worst of the rain out. Women are employed to cook for the men, and sometimes children are also used as part of the workforce to keep the machinery moving and pumping at all hours. Child poverty, malnutrition, murder, rape, and even slavery are often associated with the hard life faced by workers in the unregulated gold mining industry of this region.[16]

The gold miners work by setting up installations where mud and stone from the riverbed is shoveled into the top of a tall wooden contraption.

Generators are used to power pumps that suck water from the river to the top of the contraption, where the sludge has been deposited. The water washes the mud away from the larger stones and allows it to flow down a wooden slope that is covered in thick carpet. The carpet picks up the residues as they flow downward, capturing the gold dust, which is mixed up with the river silt. Mercury is then used to separate out the gold from the silt, and then gold is then taken to local towns, where merchants buy it at fluctuating market prices.

There is a striking resemblance between the way gold is separated from the mud that contains it and the process by which materials are classified in the lab through the rhythmical actions of sieving. Indeed, as they swirl and shake the samples of mud through the sieves, the lab technicians look, if anything, more like the romantic image of the gold panner than do the ragged-shirted laborers hoisting rubber pipes from the river to their extraction machines, or the burly men driving the enormous diggers used by the larger gold extraction operations. It is perhaps unsurprising that people we spoke to along the road wondered whether the engineers themselves were also searching for gold in their engagements with the soil. In fact, we were told in no uncertain terms that attempting to turn the process of road building into a process of gold extraction was a sackable offense within the construction company. On the other hand, tapping those who had been involved in gold mining for the stones that were left over from the mining process was seen as economically astute, and in some cases even the epitome of moral and social consciousness. In one of the conversations we had with the community relations manager of the road construction consortium in the Amazon region, he remembered an incident where the consortium's need for stone provided a means of helping out a local man who had been left in dire straits in his dealings with the gold miners who had recently passed through. The man in question was a farmer who cultivated land on the banks of the Madre de Dios River. He had been approached by gold miners who thought that the land he was cultivating was likely to be rich in gold deposits due to the river silt that was probably lying deep underneath the topsoil. With promises that they would merely take the deep-lying silt from under the fertile land and replace the topsoil once they were finished, the man agreed, for a fee, that the miners could dig up his land. But when they were finished they merely filled in the hole with the stones left over from the extraction process, leaving the former fields unusable for agriculture. When the community relations team at the engineering consortium heard about this case, they saw an opportunity. They offered to remove the stones from the man's land, and in return to bring high-quality soil from another part of the road, which they had been excavating. The man had agreed, and the story ended on a reportedly happy note, with the company in possession of much-needed stone and the man able once again to practice his agriculture.

In addition to these heartwarming stories told by the community relations team, much more hard-nosed deals were worked out whereby the consortium negotiated the ongoing purchase of stones left over by the larger gold mining companies. The stones left over from the gold mining ended up doing two jobs. First, it allowed the engineers to make the road in a way that met the design specifications, and second, it also dealt with the detritus of mining and promised to fulfil some of the road builders' environmental obligations. In terms of the mathematical calculations that determined the need for the use of stones in the road construction process, the means by which the stones had been garnered was largely irrelevant. However, extending the engineering relationship from analysis to one that also incorporated political negotiation, the ability to source materials, operated, like the math, on a contingent "as long as" philosophy. Rocks should be used "as long as" they generated a road surface that could hold up under particular conditions. Moreover, stones from gold mining were appropriate "as long as" their use did not undermine other obligations that the company had to environmental responsibility. Using stone that was the byproduct of gold mining was more economically viable than other available means of sourcing materials, and it also avoided potential upset to local residents in the Andes who on previous occasions had protested against the construction company's extraction of rocks from mountainside quarries. Conventionally, processes of political negotiation are seen as a distraction from the work of producing numbers. By emphasizing the importance of the "as long as" philosophy to both, we hope to have illustrated that, in fact, the requirement that engineers be both negotiators and mathematicians is not an aberration, but remains consistent with the pragmatic means through which road engineers approach the environments within which they work. The engineering consortium's negotiations with the gold miners illustrates that the calculative practices of engineering have the effect of making possible often surprising material reconfigurations. The specificity of the relationships produced as a result of the universal promise of mathematical calculation returns engineering practice to history, revealing modernist forms of engineering expertise to be a historically and socially specific method of engaging the unstable and processual nature of material relations. The engineer's humorous self-characterization of themselves as a tribe of "CONIRSAs," which we encountered in the previous chapter, is a perfect example of their own self-awareness of the historical specificity of their commitment to the standards and universals of engineering practice.

This would seem to contradict the distinction famously made by Claude Lévi-Strauss (1966) between the distinctive knowledge practices of the *bricoleur* and the engineer. In his classic account of the differences between mythological and scientific thought, Lévi-Strauss mobilized an opposition between *bricoleurs*, who, finding themselves in a specific material and intellectual environment, use the means at their disposal to tackle problems that present themselves, and engineers, who, by contrast, work with a more gridded understanding that isolates a specific problem and sets about finding a solution through innovation and experimentation. In the philosophy of "as if," materials are demarcated in advance according to some kind of stable and useful fictions, otherwise manifested here as so many standards and norms. Thus, we might well have come to the conclusion that we were observing a classic example of Lévi-Strauss's engineer in the process of developing specialized conceptual tools to open new horizons. However, if we extend our conception of the engineer out to their relations with materials that follow a conditional rather than subjunctive philosophy, then we see that their orientation to experimentation does not displace the sensibility or the methods of the *bricoleur*. What we wish to emphasize, from our ethnographic observations, is that the engineers we worked with were constantly aware of the provisionality of their engagement with the world. While seeking out new ways of engaging materials for productive effect, actual rather than possible solutions were constrained by established values, habits, and principles— including the availability of materials. Furthermore, the ingenuity of the engineer in such situations of constraint was a central dimension of their professional expertise, which derived as much from their capacities as *bricoleurs* as from their ability to produce and maintain distinct, abstract conceptual repertoires (Abbott 1988).

Numbers and their mathematical relations were key to this holding together of science and bricolage. We might say that in mobilizing a philosophy of "as long as," engineers were able to enact not a positive relationship with the future, as seen in the philosophy of "as if," where the road's absolute constitution could hypothetically be predicted, but rather a negative relationship with the environment, where those possibilities that could be envisaged and imagined had been mitigated as much as possible. It mattered less whether the calculations were accurate in and of themselves, and more that they produced a road that would not fall apart or contravene particular standards or regulations. Rotman's assertion as to the truth value of mathematics is helpful here:

> The question over whether a mathematical assertion, a prediction, can be said to *be* "true" (or accurate or correct) collapses into a problem about the tense of the verb. A prediction—about some determinate world for which true and false make sense—might in the future be seen to be true, but only *after* what it has foretold has come to pass; for only then, and not before, can what was *pre*dicted be dicted. Short of fulfilment, as is the condition of all but trivial mathematical cases, predictions can only be believed to be true. Mathematicians believe because they are persuaded to believe; so what is salient about mathematical assertions is not their supposed truth about some world that precedes them but instead the inconceivability of persuasively creating a world in which they are denied. (Rotman 2000, 41; emphasis in original)

We suggest that something similar holds for engineering too—engineers are working with mathematics not because it is self-evidently a manifestation of a universal truth, but because of the inconceivability of persuasively constructing an environment where such mathematical tools are denied. As long as they continue to work, they continue to be used.

Here we return to the engineer's assertion that he believed in math but not in ghosts. As we have shown in this chapter, engineers are wholly aware of the contingency of their practices. If mathematics is important to engineering it is because it is the means through which that contingency can be managed; it is the best of all possible solutions given the constraints, rather than a truth that is incontestably correct. Unlike Lévi-Strauss's caricature of the engineer who operates according to an objective problem to be solved, what we hope to have shown here is that the world is, even for engineers, not fully known in advance, and building is more than just an application of innovative abstract techniques on knowable surfaces. Mathematics provides a stable point of passage through which the relations between engineers and the substances they encounter has to pass, and engineers work with the provisional evidence that mathematics usually provides the best means available for navigating the instability of material relations in their attempts to build structures with useful social effects. However, as the statement about "belief" in math implies, mathematics, as a set of relational possibilities, does not determine how the material world will behave in any particular instance. There are always potentially more stories that could be told about the ingenuity required to order and transform the environment.

In the course of an interview about planning, knowledge, and expertise, Guido, a highway planner with many years of experience in the region, gave us the following account of a remarkable event. In 1989, Guido had assumed a leading role in the Cusco Regional Development Corporation, and one of his goals was to complete the construction of a bridge on the road to the Manu National Park in the Paucartambo region. This bridge, which had originally been commissioned in the late 1960s, had never been successfully erected. Prefabricated, it had to be levered into place with a system of rollers, pulleys, and counterweights, and the construction of a provisional iron support, which acted as a prop to hold the bridge while it was guided across the gap between its permanent supports at either end. This basic operation had been attempted several times, always in the dry season when water levels in the river were low and it was feasible to channel the flow, leaving space for the prop to be erected. Every year, just as they got to the point of launching the bridge out toward the prop, a flash flood had suddenly appeared and washed the prop away: "The river just flooded, from one side to the other—brutal! and all of a sudden." Year after year during the 1980s they had tried to get the bridge erected, but in the end the company went broke. Finally, in 1989 Guido was back on the project, and this time he took more precautions, started earlier, and everything was ready. But, just as they were about to start, the river flooded. "The prop was wobbling," he told us. "It was really dangerous." The bridge was worth over a million U.S. dollars, and if it fell they would lose everything. The chief engineer refused to continue. But the president of the corporation was an anthropologist. He was in Lima at the time, and when Guido told him what was going on he said he would fly up to Cusco the following morning. From the airport the president drove directly to Q'eros, a remote region renowned for its shamanic traditions. There, he took advice, and then he returned to Cusco and went to the market to buy all that was needed to make a payment to the earth. By dawn he was by the bridge, making the offering and saying the necessary prayers. That day the river stayed dry and the bridge was put up. As Guido remarked, "It should have been a novel!"

We asked Guido how he explained this outcome. Was it just a coincidence, or did he have another explanation? "Well," he replied, "a technical man might tell you it was a coincidence. But there are other ways of looking at it." There was already a bridge over the river at this point, but it was not suitable for the heavy loads of timber that were coming up from the forest. The old bridge could only support ten-ton loads, whereas the new one was going to support around sixty tons. On average, trucks would arrive with loads of thirty or so tons, so with the old bridge they had to stop and unload, transporting ten tons over at a time, unloading, reloading, back and forth until they were ready to set off again. A small town had grown up alongside the bridge with a thriving economy built around the services offered to the truck drivers—the loading and unloading, the sale of drinks and food, bars to get drunk in at the end of day, and accommodation to sleep in overnight. The company president had noticed this and remarked to Guido that "somebody here has made their own payments to ensure that the bridge does not get built." Guido was never sure whether his boss had been serious or whether he had been joking, but the president told him, "I brought a shaman down from Q'eros, not an ordinary one, but an *altomisayoq* [the most specialized and skilled of Andean shamans]. An ordinary shaman could never have done this, I had to bring the best, and he destroyed his rival; it worked. But, you'll see, the town will die." And so it did. The town no longer exists, and the trucks just roll by.

Chapter 4

Health and Safety and the Politics of Safe Living

Fieldwork on the Interoceanic Highway involved traveling up and down stretches of highway that had seen more than their fair share of accidents and fatalities. Crosses and roadside shrines marked the sites where people had lost their lives. Small chapels had been erected in notorious spots by travelers seeking divine protection. The road between Puerto Maldonado and Mazuko is renowned for being particularly dangerous. Drivers have to navigate a series of winding curves and deep ravines, where rickety wooden bridges are laid over terrifying precipices. When we reached the end of one of these bridges, colloquially known as Puente del Diablo (Devil's Bridge), we found a well-maintained concrete building, which was painted blue and had a red, corrugated iron roof. Behind a locked metal grate we could see seven burning white candles, several jars of artificial flowers, and a statue of El Señor de la Cumbre the Christ figure to whom the shrine was dedicated.

Later on we discovered that the construction company was faced with a dilemma here. Once the road was widened there would be no room for the shrine, and the company hoped there would be no further need for the

Figure 9. Roadside shrine at the Puente del Diablo

protection it offered. Yet the shrine registered past events as well as offering protection in the future. It was a memorial site, a place dedicated to those who had already died. It was decided that it should be moved, rather than removed. Everybody knew somebody who had died on the road, through crashes, landslides, fallen bridges, and collapsed surfaces, and these small chapels served as collective sites of contemplation and drew responses from long-distance travelers who would cross themselves or mutter a prayer as they passed by.

Benedicto Kalinowski had traveled the Interoceanic Highway more than most over the years, and understood the dangers it posed:

> Imagine, you have your husband and your children, and you want to go to the jungle to work? Let's go and work there! You go via this bad road. You arrive eventually almost without any desire to be there. But in the end you have arrived. First you make a little shack out of leaves and start to work. But you always have to go to back to Cusco to get the right equipment to grow sugar, and you have to bring all your food. Your husband, let's say he goes to collect some food to help you continue working there. On the road there is an accident and your husband dies. Now you are a woman living alone with two children or a child, and this is the problem. From there comes the failure. And it is always like this. It will always fail if you don't have a good road. There is no good road here, it's just a path—not even that. A road is another thing. This is not a road, it is suicide. You have to go to die and nothing else on this road.

Kalinowksi's sentiments were echoed by local people along both the roads we studied. The need to mitigate the dangers of road travel was at the forefront of campaigns for infrastructural development. The construction consortium was well aware of the importance of providing new, safer surfaces, and it frequently presented the value of its work by displaying before and after shots, images that contrasted the misery of past road travel—overturned trucks, routes blocked by landslides and boulders, treacherous muddy or inundated surfaces, broken bridges and rivers in full flood, passengers stranded—with the clear, smooth surface of newly finished stretches.

However, the company and local people alike were aware that the new roads brought new dangers. Slow traffic had protected the lives of children, animals, and vulnerable adults. Changed routes also meant that the roads became unfamiliar. Beyond the dangers associated with the hugely increased speeds, there were stories of drivers missing the curves or failing to align their vehicles with new bridges. The construction process itself introduced a new sense of vulnerability. Early on in our fieldwork we were approached by the authorities of one rural town who hoped that we might have some insights about how to best handle the psychological and social dangers associated with the arrival of the workforce. They feared that the presence of so many single men would lead to theft, unwanted pregnancies, social unrest, and unwelcome disturbance. As construction got under way local people were dismayed by the noise, the dust, and huge machinery operating day and night in areas that had always before been quiet and still.

Reflecting on the contrasting responses from the U.S. authorities to the devastation of Hurricane Katrina and Hurricane Sandy, Stephen Collier has drawn attention to two quite distinct understandings of vulnerability. Collier argues that in the case of Hurricane Katrina it was "social vulnerability," the exclusions and precarious living conditions of the poor and marginalized populations of New Orleans, that came to the fore in the media and in analyses more generally about why the city had been so devastated by the floods. The social vulnerability response elicits "infrastructure" as the solution. Investment in new infrastructures mitigates such vulnerability via a material politics of inclusion and connectivity. In New York, the case for social vulnerability was less clear, and pundits focused more squarely on the vulnerability of the systems themselves. Infrastructures were not lacking, but they were at risk, their systemic support for life-as-usual revealing vulnerabilities that were not otherwise routinely considered. Elsewhere, Collier and Andrew Lakoff have argued that concerns about critical infrastructures and the vulnerabilities they have come to represent go hand in hand with expanding regimes of securitization—that is, public investment in the protection of infrastructures in the name of the common good (Collier and Lakoff 2008). These concerns with infrastructural security are often controversial, particularly when they involve investment choices that divert resources from other modes of welfare. At a time when many public infrastructures are in effect run through public–private finance initiatives, the securitization of infrastructures appears to many people as state protection of private interests. In a more general sense, others argue that the contemporary turn to security is a depoliticizing move that replaces concerns with rights, inequality, and justice with what Gourevitch (2010) calls a "politics of fear," which is concerned only with the protection of the individual and conjuring the uncontainable dangers of terrorism or environmental catastrophe.[2] In this chapter, we bring to debates around infrastructural security a detailed ethnographic look at the complex and varied experiences of vulnerability associated with particular infrastructural forms. Such an approach allows us to rescale the terms of the debate, such that the relationship between the risks of systemic failure and the need to build strong relations of trust and mutuality can be properly explored. In other words, attention to infrastructural politics requires an approach that works across scales and does not allow the abstraction of systemic relations to be disconnected from the specificities of interpersonal relations. Our aim in this chapter is thus to trace both the possibilities and the consequences of the security and health and safety measures that accompanied the construction process. In particular, we explore the contrasts suggested by Collier's work between a past-oriented temporal framing, in which vulnerability emerges from specific prior relations of neglect and where risk can be averted by building strong relations of trust and social resilience, and a future-oriented approach, which seeks to manage risk at a distance by modeling the consequences of transformational processes, in anticipation of potential and unforeseen dangers.

The First Death

The construction consortium took health and safety seriously and was committed to security and safe work in many different ways. The construction camps isolated the workforce from adjoining settlements with high perimeter fences and carefully patrolled entry points. The logistics managers carefully controlled the movements of all machinery, and all the materials were kept under close surveillance. Although the consortium had no direct authority over other road users, it did control the opening and closure of the road itself, and the central government had ordered local police to enforce the road closures in the name of public safety. In chapter 6 we look in more detail at the contradictions that arise when public security officials are deployed to protect the private realization of public works. For now, however, we focus more directly on the regimes of health and safety surrounding the construction process itself.

Long before the construction camps were built there were engineers out on the roads, drawing up plans, looking for sites from which to extract or dispose of materials, negotiating with local municipalities over labor agreements, and working to gather the information needed for the final design of the route. They needed to be housed and fed, and local restaurants and small hotels found themselves subjected to health and safety standards well beyond anything that they had previously been used to. Only those establishments that met the criteria imposed by the consortium received the business that it could bring. Standards improved dramatically. Environmental protection was also of concern. Environmental officers embarked on a range of educational initiatives in an effort to improve the habits of local people, urging them to dispose of their rubbish responsibly, not to wash vehicles in the rivers, to protect water sources, and so forth. Sanitation was also a concern. Portable toilets were set up along the route for the use of the workforce. Inside the camps, life was ordered, hygienic, and controlled. And as far as the environmental officers were able, these new standards of order and improvement were rolled out to communities as part and parcel of what came along with the road itself.

But road construction is a dangerous business, and it was not long into our fieldwork before we heard about the first construction death. News traveled fast, and by the evening of the day of the accident the town of Ocongate (several hours' drive away from the accident site) was buzzing with news of the fatality. No one knew quite what had happened, but everyone wanted to talk about it. A few days later we drove past the spot where it had occurred. Two wooden crosses were hammered into the ground—one provided by the construction company and one by the relatives of the deceased man. It was a tragic accident. The man had been working behind a bulldozer, and it seemed he had tripped or somehow been distracted, and the machine had driven right over him and crushed him. The driver had instantly been dismissed.

At the entrance to one of the construction camps a large sign declared: "our target is zero accidents" (*nuestra meta es zero accidentes*). In retrospect, the target seemed to herald rather than prevent the occurrence of such fatalities, and it was curious to note that the sign was neither taken down nor changed after the first death. The fact that the actual death did not displace the target of zero deaths drew our attention to the particular way in which human life as a universal value played a central role in the health and safety regimes of the construction consortium. This is not to say that the death was ignored. On the contrary, the event was taken very seriously and explicitly

Figure 10. "Our aim is zero accidents"—health and safety at the Iberia camp

incorporated into the health and safety training that all new employees of the company were required to attend. As we explore in more detail below, these training sessions present health and safety measures as uncontroversial, universal, and necessary. The primary importance of human life over all other considerations is posed as a given, and both the company and workers are expected to orient themselves to the objective of protecting their own lives and the lives of others.

However, we found that the commitment of ensuring health and safety for all was under constant threat of compromise from all the other relations that are necessary for the successful construction of a road. The apparently ubiquitous concern for safety above all other preoccupations was continuously undermined by the practices of both the consortium managers and its employees. A concern for human life thus appeared to form the basis of health and safety regulation and to exceed it as people became preoccupied with dimensions of life that lay outside the remit of legislative control. In what follows we look at the way in which human life is appealed to and established as a universal principle of safe working and of safe living. We also explore how the production of "human life and death" as a universal preoccupation relies on the mundane enactments of formalized processes and regulatory standards. Health and safety legislation places the question of human life (and death) at the heart of projects of infrastructural development. The training of the workforce is oriented to the question of protecting that life, and the consequences of ignoring rules and regulations oriented toward this end are portrayed as dire and traumatic. Health and safety regimes in practice emerge through particular material relations with equipment, environments, and other people. Thus, even though everyone agreed to protect life in the abstract, no one could possibly find a way of living in the abstract. Health and safety was simultaneously produced as a technical procedure of bureaucratic control and as an affectively charged site where fundamental issues over what it means to participate as a responsible member of a social collective were brought to the fore and actions that appeared to undermine the protection of life were scrutinized and punished. By establishing a universal version of life lived in the abstract, health and safety thus cleared the ground wherein a space for the respecification of the relations appropriate to life as a universal good were to be established and contested. In this respect, health and safety can be cast as a highly politicized sphere, where lines of difference, boundaries between insiders and outsiders, and practices of incorporation or exclusion are reinforced.

Induction into Safe Working

Shortly after we heard about the accident, the opportunity arose to attend an induction session for new workers. The room was stuffy and filled with employees, and we had already sat through several presentations when the health and safety officer, an engineer known by employees as Revaleta, came in, flanked by two colleagues. The friendly, relaxed tone of the previous sessions evaporated as Revaleta dramatically flicked off the lights and demanded people's absolute attention. The room was now only lit by the glow of the PowerPoint projector. Revaleta's colleagues patrolled the room, watching us, ensuring we remained attentive, clapping their hands above the head of anyone who appeared to be losing attention or drifting off to sleep in the soporific quiet of the darkened classroom. The talk on accident prevention was quite abstract. Accidents, Revaleta asserted, need never happen. There is always a reason for accidents, and they can always be traced back to human error. He wanted to make sure that everybody understood that point: there could never be an excuse; there were no supernatural forces; there was no misfortune or bad luck; there was only failure to attend to the correct procedures.

The image of a man lying dead on the ground, his head in a pool of blood, flashed up on the screen. Somewhat shockingly, it turned out this was the man who had died in the accident earlier in the week. Revaleta's theatrical gestures reached a peak as he declared that this death was avoidable. Neither the victim nor the driver had been attentive enough to what was going on around them. Had they been, the man would not be dead. There was no sympathy expressed, no acknowledgment of the dangers and difficulties of the work. The target of "zero accidents" was reiterated as a key objective that drove Revaleta's passion to ensure that the construction process was safe. Revaleta emphasized that it was his job to make sure that everyone was signed up to this common goal. Not only was safety important, but for Revaleta and the company alike, including workers at all levels from the most senior engineer to the lowliest worker, safety was an absolute priority. And anybody who wanted to become, and remain, a member of the "family" of the construction company had to embrace security as his or her principal objective.

Revaleta knew that those sitting listening to him in the room were nervous about being singled out, or picked on, but he also knew that they were not likely to agree with his unforgiving attitude, and the implication that those who had accidents were always in some way to blame. The purpose of the talk was to explain the logic of personal responsibility. To do this Revaleta defined his terms, distinguishing between dangers, risks, accidents, and incidents. Dangers (*peligros*), he told us, are things that could potentially cause accidents. Construction sites are full of dangers, and workers have to be aware of them. Then there are risks (*riesgos*), which can be assessed and minimized. Risk is a relational condition, such that $R = P + C$ (Risk = Probability + Consequence). That is, the level of risk (R), expressed on a scale of 1–10, is directly related to the combined value of probability (P) (the likelihood of occurrence of a scale of 1–10) and consequence (C) (the potential harm caused on a scale of 1–10). An accident (*accidente*) occurs when risk is actualized as personal injury, material harm, or interruption of the construction process. An incident (*incidente*), on the other hand, is when risk is actualized in a way that does not involve (but could have involved) personal injury, material harm, or the interruption of the construction process. Revaleta gave the example of dropping a tool that just misses causing injury to a fellow worker. An "incident" in Revaleta's algebra of safety is as grave an event as an "accident," as it indicates the same level of irresponsibility on the part of those involved, and such irresponsibility cannot be tolerated on a worksite committed to safety.

The notion of unpredictability was of no interest to Revaleta. His responsibility was to specify dangers and to teach people to manage them so that working conditions were not experienced as unpredictable. Dangers were dealt with by introducing a series of norms, such as bans on drinking, requirements to use safety equipment in particular spaces and circumstances, procedures for clear communication between foremen and workers, and stipulations about hours of work that would prevent exposure to the dangers of fatigue. Dangers could and should be foreseen, risk could and should be minimized, and accidents and incidents should never happen. The target for deaths, accidents, and incidents remained at zero, for the rules were absolute, and the relationship between cause and effect constant. Under this regime of safe working, workers were thus required to conduct themselves in particular ways and to participate in a series of material practices and gestures, all of which assumed (and thus enacted) a commitment to this line of reasoning, ultimately underpinned by the stated objective of protecting human life. The logic was harsh. To be involved in an accident or an incident of any kind implied a contravention of this logic, and by extension it demonstrated an active lack of care for human life.

Revaleta acknowledged that rules were flouted, and that accidents happened. But, again, the causes came with no excuse for those involved. They included lack of knowledge, inadequate psychology, problematic attitudes, inadequate leadership, and a deficiency in the training of engineers. Ultimately, the goal of safe working required individual workers to take personal responsibility both for their own safety and for the safety of others by acknowledging danger and minimizing risk.

If we limit ourselves to the discursive dimensions of Revaleta's exhortations to the new recruits, it is easy to see how his health and safety regime might be interpreted as another example of the politics of fear or the latest incarnation of the risk society.[3] Revaleta's call for the new recruits to rethink their actions as potentially destructive of life depended on the erasure of other terms within which these actions might be understood or interpreted. Risk was made to stand as the unifying dimension, which could supersede the varieties of meaning that people might incidentally layer onto their objective participation in any particular activity. The assertion of risk as a fundamental equation underlying all actions had the effect of making alternative acts of interpretation appear as an obfuscating layer, as false excuses that needed to be stripped away and replaced with a universally applicable set of rules.

Yet, as Tsing (2005) reminds us, the act of generalizing toward the universal is "an aspiration, an always unfinished achievement, rather than the confirmation of a pre-formed law" (Tsing 2005a, 7). Approaching universals as emergent processes, rather than as fundamental beliefs, radically transforms the possibilities for thinking about the politics of infrastructural projects. It requires us to move away from a concern with mapping opposing beliefs and to move toward the question of how to account for the emergence and stabilization of so many positions in the shifting terrain of interests and responsibilities that infrastructural projects entail. It requires us to move from the question of who believes what to how a political terrain is constituted out of material practices of interaction.[4]

In what follows we focus not simply on the discursive form through which health and safety procedures were justified but on the various ways through which safe living was achieved in practice. In road construction, health and safety practice involves a wide range of material supports alongside the regulatory frameworks. From road signs, which offer orientation and warning to drivers, to a manual of health and safety rules, which is read out to work teams on a daily basis, from speed limits and traffic humps to the presence of Engineer Revaleta himself, the project of health and safety required the participation of any number of "assistants," to use Jörg Beckmann's term (Beckmann 2004).[5] These assistants, or supportive devices, actualize the stipulations of the health and safety code, taking on the weight of obligation and the commitment to protect life. Ethnographically, they also offer us a way of apprehending alternatives as, unlike the abstract commitment to human life, the mundane practices through which these devices enroll workers into particular modes of safe working. These supportive devices also raise questions about a variable and discriminating regime of protection that does not, in practice, offer equivalent safety to all.

Enacting Health and Safety

The construction process is regulated by norms of engineering, by state law, and by the enactment of the company's responsibility to protect against potential harm both to and from the social and material environments through which the road passes. In the context of the engineering project the engineering consortium devises and enacts a code of corporate care that works across all dimensions of the project. Ultimately, the project of building a road is itself a process that aims to produce a systematic stabilization from an unstable social and material world. It represents an infrastructural response to social vulnerability. At the same time, the code of corporate care is itself a regime of transformation and reduction of risk; it is a systemic mode of protection that secures its own resilience in anticipation of many different kinds of danger. The divisions of the company responsible for human resources and for environmental and community relations are all engaged in processes of transformation and protection. Nonetheless, in spite of continuous attempts to standardize practices through the enforcement of universal rules and regulations, we found that the practice of safe living repeatedly brought the rules into question.

On signing a contract of employment with CONIRSA, each of the workers employed on the Interoceanic Highway project was provided with a range of safety equipment, from steel-capped boots to reflective vests, gloves, safety glasses, and face masks. The equipment provided varied depending on the particular demands of the job they were employed to do, but one piece of equipment provided to all members of the construction process, from the manual workers to the senior manager, was a hard hat. The hats were color coded according to the occupational classification of the person wearing them: blue hats were worn by qualified workers such as drivers; red hats by security; white hats by the engineers; green hats by the foremen; and yellow hats by the *obreros*, or nonqualified workers. Although the colour coding had certain social effects, such as demarcating a hierarchy of roles and responsibilities, what we were equally interested in was the ways in which the hard hat itself took on an importance beyond this significatory color coding.

When traveling in a car owned by the consortium we were told we had to have hard hats at the ready in case Revaleta should appear. All employees were required to wear their hard hats when traveling in company cars. On one occasion Penny even had to get out of a car she was traveling in and pretend she had nothing to do with the other passengers, because Revaleta had been spotted nearby and the driver she was traveling with had forgotten to give her a hard hat to wear. The driver explained that he might lose his job, as he was responsible for the safety of all those in his car. A little while later, when the danger represented by Revaleta had passed and Penny had been allowed back into the car, the other passengers spotted someone in another company car without a hat on. When they pointed it out to the driver, he laughed and said, "he must be Revaleta's cousin!" The suggestion was that the kinship relation would offer that driver a degree of job security that others could only acquire by strict adherence to the safety code.

Throughout the induction talk Revaleta had stressed how infringement of the rules would lead to immediate dismissal, and he was widely feared within the company because he had a reputation for carrying out this threat. He had been known to send people home without pay for as minor an offense as failing to eat breakfast—an infringement of the requirement to ensure that you turn up for work alert and strong, capable of safe working. Low blood sugar or hunger can lead to lack of concentration, which increases the risk of accidents.

At the same time the jokes in the car about Revaleta's cousin point to other ways of being safe on the job. Knowing the right people and being in the right relations allows some acceptance of the fact that most people do not wear protective headgear inside the cabs of modern vehicles. The wearing of such helmets is not prescribed by law and was widely seen as an excessive precaution, perhaps more indicative of policing workers and monitoring their willingness to accept the discipline required to show themselves as responsible employees. To act in accordance with the rules shows that you can be trusted and be treated as a member of the CONIRSA family. The irony of the dispensation for existing relatives was not lost on the workers.

The inherent ambiguities in the tension between modes of vulnerability and the consequent understandings of safe living were not played out only by those looking for ways around the inconvenience of the constant self-monitoring required by Revaleta's regime. These ambiguities also emerge in the wake of corporate gestures to build the sense of "family" on which, in a different register, the health and well-being of the project also relied. On Peruvian Independence Day the construction camp, which was normally presented and managed as a space of inherent danger, was suddenly declared a space of celebration. The company was hosting a party to celebrate this national festival, and normal safety procedures were suspended. Workers were told that they could wear their own clothing that day, that they could drink beer, dance, and joke, and leave their hard hats at home. The camp was physically transformed as bunting was strung up in the national colors, and a band was contracted to ensure the right festive atmosphere. There was a barbeque and ice cream, and the low-paid Peruvian workers were encouraged to mix with the Brazilian managers as all stood side by side with their hands on their hearts to sing the national anthem of Peru, salute the Peruvian flag, and join the rest of the country in a celebratory display of national unity.

The conditions of safe living on Independence Day were clearly quite confusing. Employees were told that although they were not expected to turn up for work that day, they were expected to attend the party. And, although it was a celebration of national unity, they were also told that they were not permitted to bring their families into the camp. Despite having been told that they should come to the event in nonwork clothes, when the day came many workers turned up to the party in uniform, with their hard hats under their arms. It was not so easy to suspend the "grid of intelligibility" (Foucault 1978, 93) through which hard hats and sober living had gained the power to prevent accidents and provide job security. For those who were constantly concerned about whether they would lose their jobs, it was of huge importance not to step out of line. Was this party work or not? Was the camp a space of danger, or a space of celebration? Were these mutually exclusive? Were employees free to eat, drink, and dance without fear of retribution, or was their behavior still subject to surveillance by the foremen and managers who were also at the party? The uncertain dimensions of the transparent and universal code had come into view, and many workers turned up at the party in their full safety attire.

The seemingly arbitrary way in which rules can be applied and removed, and the ways in which this oscillation creates spaces in which people make mistakes about how to behave at different moments in time, often makes it even more difficult to enforce these rules when people decide for themselves that their circumstances do not warrant their application, or at least not in the ways intended. A recurrent worry for the health and safety team were people employed as signalers on the road who kept removing the rubber face masks intended to protect them from the dust because the masks were hot and sticky. Having spent all their lives traveling along dusty roads without face masks, these workers did not prioritize the dangers to health over the discomfort of standing overclothed in the sweltering heat. In these circumstances, the protection seemed excessive. It impinged on their concentration and raised the question of what mattered more—an abstract and uncertain protection for a specific aspect of long-term health, or an immediate sense of basic comfort that allowed them to get on with the job. There are no clear answers here, but the universal code leaves no room for discussion. As with the requirement to wear hard hats in company vehicles, workers suspected that the capacity of the masks to protect against dust was rather less important than their capacity to protect against dismissal. Here, we suggest that the ambiguity of the relationship between the abstract health and safety code and the specific practices through which it is enacted produces the space in which people then take responsibility for themselves in ways that exceed the terms of the relationship they agreed to in signing their contract with the company. Indeed, in the daily practice of construction work, the enactment of responsibility is as likely to be the source of apparent rule flouting as it is the basis on which to ensure that rules are strictly followed.

The Deflection of Harm through the Building of Trust

We return now to the accident with which we opened this chapter. The ambivalences that cling to incidences of harm emerge in the tensions that exist between a systematic and logical health and safety code and the chaotic, uncoded, and incomprehensible "outside" that such a code posits as its external other. Our analysis thus far has suggested that putting formal health and safety regulations into practice opens up a new set of relations that are neither internal to the rules nor outside of them. Moving away from the idea that formal codes are simply imposed, we look now in more detail at how people negotiated the potential for harm in road construction projects.

People employed by the construction company were saddened but not surprised by the news that a worker had been killed. This was, they said, the first in what was bound to be a series of fatalities. The accident happened in a landscape imbued with the presence of powerful spiritual forces, and the death was seen by many as retribution for the suffering of the land as it was gouged out, displaced, and removed in the process of road construction. A driver told us in confidence, when the engineers were out of earshot, that the earth had grabbed hold of the man's legs and prevented him from getting out of the way as the machine rolled over him and crushed him. There was a recognition that these kinds of occurrences happen all the time. Since the accident there had been reports of other events. There were rumors that some of the dead man's coworkers had begun to be haunted by his spirit, while others had fallen sick inexplicably. A bulldozer was said to have mysteriously crossed in front of one of the night drivers, almost causing another accident. We were told that the death would be just the first of many. The driver calculated that a project such as this would likely bring eight or nine fatalities.

How then did people deal with these kinds of threats to workers' safety? How did they approach the combined inevitability and unpredictability of harm that this project would bring in the process of delivering them a better, safer future? One means of mitigating the potential anger of the earth deities was to ensure that proper ritual payments were made to these spiritual forces before the earth was tampered with. The people who described the accident as the outcome of unpredictable forces of retribution were used to making such payments at the beginning of the agricultural cycle, in the hope of an abundant harvest; at the start of any construction project or business venture; and even when setting out on a journey.[6] There was recognition that a similar ritual would be appropriate in relation to the process of road building.

Aware of this ritual practice in the region the road was to pass through, the community relations team at the engineering company had in fact performed a ceremony before the construction began. They made ritual payments to the earth in the form of the Andean *despacho*. In the Andes people make offerings to mountain spirits and earth deities in an attempt to draw these animate forces into relationships of exchange for mutual benefit. Huge care is put into the details of such offerings, which have to be performed accurately to avoid antagonizing the powers that animate the Andean landscape. These powers have the capacity to ruin harvests, kill livestock, and send illness and death. They also have a capacity for benevolence and can bring good fortune to a lucky few. Forging a relationship with mountain spirits in this way is, however, explicitly without guarantee. Humans can petition, but they do not control.[7]

The performance of an Andean *despacho* by the engineering company was therefore highly ambiguous. The authenticity of the practice was dismissed by some observers as impossibly naïve or deeply cynical, carried out as it was by a consortium whose managers had no personal investment in the land and no belief in the importance of maintaining good relations with the land as a powerful and sentient life force. However, they did have an interest in calming the workforce. Thus, the engineering consortium decided to hold a Catholic mass for the worker who had been killed. As noted above, the death had led to a proliferation of hauntings and ghostly threats, and workers were refusing to come to the site where the man had died. Anxious to put an end to the contagious effects of the first death, the company managers recognized that the apprehensions of their workers needed to be engaged, even if the managers' primary motivation was to calm the workers rather than the earth itself. In the mode of corporate and social responsibility, the company was used to exercising a duty of care in a way that was recognized as transparent and reproducible. In contrast, here they were drawn into engagements that required a sensitivity and attention to dimensions of social relations that were specific to the places through which the road was to pass. The duty of care was produced here not as a generic systematizing logic, but rather as a specific relational capacity. These relations of care nonetheless operated within powerful, if not fixed, parameters of their own. The company policy was certainly to be respectful of local beliefs, but as in so many other postcolonial and development "encounters," such culturalist accounts of difference have potentially discriminatory effects. The suggestion that different people simply "believe" different things masks other potentially more volatile differentiating practices, such as the capacity to attribute value, claim ownership, and express concern in a way that is recognized as valid.[8]

The differentiation of practices is very evident if we look at the relationship between the version of construction organized through concerns with health and safety and those manual construction projects that took place along the road and lay outside the remit of the company. Most of the workers in the induction class that we described above came from the local area, due to a legal stipulation imposed on the consortium that, whenever possible, it must employ labor from the locality through which the road was passing. These workers were accustomed to an Andean mode of collaborative labor through which construction projects in local communities were usually organized. Just that week, we had been in a community near the engineering camp, watching a group of people take down and reassemble two adobe houses that were in the way of a small track they had wanted to straighten and widen. The community work party was organized precisely around a kind of intense sociality that Revaleta had explicitly outlawed, as amounting almost to conspiracy to murder, under the health and safety code, where personal responsibility remained the bedrock of a safe future. Revaleta had explicitly disallowed drinking, joking, smoking, and even laughing as potential distractions that could transform danger into risk and ultimately accident, yet these were obligatory activities in Andean work parties. In these parties, the source of potential harm was not to be found in the equations that informed Revaleta's systemic approach to risk reduction, but rather in the unspecified dangers of inhabiting a world without sufficient or proper relations of care and trust. For the work party, personal vulnerability was addressed by ensuring adequate sociality with fellow workers. This required people to do things together, to drink and smoke and dance, to generate the energy and the comradeship needed to prevent the collaboration from falling apart. Far from being an attempt at control as the primary means of mitigating harm, the ways in which these work parties dealt with the dangers of an uncertain future was through the production of energetic relations. In some contexts, passing out from the excesses of alcohol consumption is taken as a sign of trust and a job properly celebrated.[9]

Here the practices oriented to ensuring the ongoing "safety" of construction required the mobilization of alcohol, coca leaves, ritual payments to telluric forces, and ties of *compadrazgo* (godparenthood) and kinship. They also involved the acknowledgement of particular skills and capacities that marked particular men as "specialists" (*maestros*) who would orient proceedings through direct involvement, taking the lead, performing the most difficult tasks, and showing others by example. Although at first glance the forces of regulation that we find in these work parties appear incommensurable with those of health and safety, attention to the way in which these parties engage harm highlights that they too are organized through hierarchical relations, and that they also mobilize extended material support to bring about their regulatory effect. If there is a difference in the way in which these work parties deal with the anticipation of danger, it is in the way in which regulation does not inhabit a space outside everyday social practices in the form of explicit rules. Rather, regulation emerges out of the ongoing relations between people and things in the course of everyday life. The management of risk in these work parties is not a technical exercise that can be imposed on social relations, but the very stuff of relating itself.

The Resurgence of Difference

Nonetheless, the codes of corporate care, of which health and safety regulations are one part, rest precisely on the assertion of a distinction between technical devices and social practices. The requirement to stabilize universal principles, and to engender forms of regulation that will ensure a robust social response to universalizing equations, has the effect of making these rules appear as devices that offer protection in and of themselves. Health and safety codes appear to enable the construction companies to detach the technical from the social, and to proceed on that basis. Within this framing, the degree to which the companies attend to social issues emerges as an added benefit, a gesture of goodwill, an agreement that enhances a bid and adds value to the core expertise of a company. And yet, as we suggested above, these codes are notoriously unstable with respect to projects of social ordering, precisely because those relations and affective forces that are carefully put to one side are a central means by which the universals and codes of conduct are themselves reproduced.

We were fascinated to observe how those responsible for health and safety in the construction process tackled this dilemma of the destabilizing gap between health and safety codes and social practice. Implicit in Revaleta's evangelical selling of the health and safety code, and in spite of the list of means of ensuring that it was followed, was the recognition that ultimately these rules were going to be circumvented. Revaleta knew that his drivers would take shortcuts down the mountainside, cutting off the curves in the tortuous bends by veering down crumbling gullies. He expected that workers would show up to work drunk, ignore the rules about wearing their hard hat in their vehicles, break the speed limits he had imposed in the name of health and safety, steal the road markers with their reflective strips, and forget to wear their face masks or florescent vests. Although this was a frustration for health and safety officers such as Revaleta, it was perhaps more interestingly a cause for the production of all kinds of creative responses by the health and safety team to draw people into the project of safe living.

Of all the many people we spent time with on the road construction project, none were as obsessive or as overtly passionate about their work as the health and safety officers. This attitude was one they openly talked about and took pride in. They loved their work, and they were fascinated by their potential to draw others into practices of safe living. Education was central, but education worked through allure, passion, and affect. We saw in the previous example of the health and safety induction how Revaleta worked on his audience with a deep sense of theatricality, evoking the insecurities that seem to require health and safety codes to ensure the conduct of safe living. Others worked via different relational possibilities. We were particularly captivated by a health and safety engineer who taught inductees by drawing his students into dramatic scenarios where they had to learn to help each other to survive. He was fascinated by his job and wanted to teach by contagious enthusiasm. The challenge was to conjure scenarios that his students could invest in emotionally, so that they would be motivated to learn the skills they needed to be able to work safely. The skills were not universalizing codes but principles of interdependence and a capacity to improvise. For example, he taught them how to make stretchers from poles and a shirt and stressed the importance of being able to respond to accidents in remote sites. We were particularly impressed by his passion for road signage. He wanted his road to have the best signage of any in Peru. Such signs are, of necessity, standard forms—themselves symbolic codes designed to disambiguate potential hazards such as bends and steep inclines, or to offer clear instructions as to speed limits. Nevertheless, his versions of these signs were hand drawn, lovingly painted, and carefully placed. He was particularly happy about the reflectors that he had fixed to bridges and curves to alert drivers to dangerous drops. These small devices had succeeded in capturing local interest to the extent that people were said to be taking special trips to see "the lights," sometimes traveling up to thirteen kilometers for the pleasure. Their popularity was such that they were also being stolen; while disappointing in some respects, this was also clear evidence that they "worked," were noticed, and were mobilizing people.

The universalizing practices that we have referred to in this chapter are usually considered the basis on which the logical practices of pursuing infrastructural solutions to security problems are enacted. However, the production and mobilization of universals in the pursuit of an infrastructural response to danger and harm requires participation in the affective fields of the social practices that they are designed to sidestep. In the social world all abstractions are unstable as they continually reenter the dynamics of exchange and circulation that render them unpredictable. In these circumstances, the options for safe living multiply. At the same time, and by extension, the material artifacts that assist this process provoke a flouting of the rules. Thus, while Revaleta positions himself as central to the project of health and safety by extending his panoptic gaze in order to engender a principle of self-discipline in his workers, he also invites a response. Driving down "illegal" shortcuts [*cortes*], the drivers of the consortium's four-by-fours tell us of Revaleta's warning to them: "remember, even the rocks are watching you!" Nonetheless, in spite of Revaleta's warnings, they remain skeptical of Revaleta's capacity to see what they are doing. These workers are simultaneously able to transfer their trust in Revaleta's rules and regulations into a trust in the capacities of the vehicle to perform the health and safety measures built into offroad vehicles as they drive their vehicles over apparently precipitous drops, even in the black of night. At the same time, there is a frisson of excitement about breaking Revaleta's rules, and possibly even wonder at the ability of the vehicle to take them to the limits of acceptable practice. Bored, away from home for long periods of time, in environments that offer them little by way of habitual pastimes beyond the television and a football pitch—the construction site and its surroundings are environments made for practical jokes, seductions, and the dangerous hide and seek with Revaleta and those who became his eyes and ears.

These examples are intended to demonstrate the importance of looking at the specificity of the relationships in which universalizing codes come into being, the risks they make visible, and the limitations or slippages that are produced when these regulatory devices are imposed, however strictly they are applied. The very rules on which infrastructure projects are launched as responses to danger and harm can become relevant only to the extent that they engage with existing modes of safe living and the established expectations of others. Even though all the security personnel that we came across in the company had a sense that they knew better, and saw local people as in need of education, the means by which they enacted and produced the health and safety code necessarily required an openness to other possibilities or modes of relating.

By paying attention to codification practices of different kinds, we find that security experts do not rule through the authoritarian imposition of a formalized health and safety code. Rather, the regulatory practices that are so central to infrastructure projects are imbued with affect, such that they operate to ensure that workers engage specific principles and mechanisms of safe living. As the product of these relations, the health and safety practices work to redraw distinctions between people operating in different domains and with different preoccupations. By identifying themselves as providers of health and safety, a manager or an expert can remain utterly committed to ensuring safe living in the fields of relations for which they have responsibility, while at the same time deflecting responsibility for anything that happens onto individual workers. At the same time, workers, who do not necessarily claim to be actively engaged in reproducing the regulatory environment, can draw attention to the contingent circumstances that are constantly in play and shape how people act in particular circumstances. Nonetheless, if these actions circumvent the relations necessary for the reproduction of the more formal system, they risk being delegitimated as either irrelevant or ignorant.

Conclusion

We have used ethnographic material to explore the regulatory devices that we found people mobilizing as ways of dealing with the anticipation of harm in road construction. Our primary aim has been to draw out the differences between the universalizing pretensions of the health and safety code and the extensive social relations through which they are enacted. Our purpose in making this analytical distinction has been to help us identify with more precision how the formal codes, which are mobilized in establishing infrastructure projects as responses to vulnerability and risk, orient practice in environments that, for some, become increasingly uncertain in the face of the transformations entailed in infrastructural development projects. We also have been concerned to find ways of analyzing the contemporary politics of difference as articulated and experienced under conditions of development and infrastructural intervention (a topic we take up in greater detail in chapter 6). In this respect we have shown that an attention to the practice of producing and sustaining universal values provides a fruitful analytical approach. It is clear that risk equations are far from neutral interventions in social situations. They configure both the mode of vulnerability and the related assumptions of appropriate action. In this view, a preoccupation with risk is less of a cultural or ideological baseline from which political effects emerge and more a way of actualizing a potential problem, such that unspecified future events can be both prefigured and acted on. As we have shown, this is done in various ways. Some of these could be seen as simply cynical responses to cultural incommensurability, such as the enactment of the offerings to the land by the engineering company. Instead, however, by attuning ourselves to the copresence of different techniques that aim to bring about a safer future under conditions of uncertainty, we are able to describe not a clash between closed worlds that imitate or impinge on one another, but rather a set of contingent and constraining practices that can operate as explanations for the dynamics of contradiction and confusion, as well as those of more formal conduct and control.

Hopefully, then, we have come a little closer to evoking a politics of differentiation that emerges through engagements with the danger produced by projects of infrastructural development. In this sense, we would distance ourselves from a diagnosis of a politics of fear whereby "the political system" imposes a depoliticizing regime of technicized concern with systemic vulnerability, ignoring the historical concerns that underpin the more overtly political understanding of social vulnerability. Our concern is that this dichotomy fails to address the ways in which infrastructure both separates and unites state, corporate, and local agencies. Our focus on the tensions between social and systemic vulnerabilities shows instead that dangers appear in different forms, at different times, and for different people. Rather than set up a rational corporate health and safety regime against the nonrational assumptions of a local population (as Revaleta did in his theatrical delivery of the health and safety induction), we set out to produce a more integrated account of how people address the dangers and uncertainties of projects of infrastructural transformation. These include Revaleta's calls for personal responsibility, the ritual payments to the earth, the evocation of wonder in road signs, and the trust placed in mechanical vehicles. In this way we are not forced to collapse the differences that continue to provoke and engender both suspicion and wonder about the dynamics of change and the possible futures that dangerous but desired technologies like roads might bring into being.

Chapter 5

CORRUPTION AND PUBLIC WORKS

The whiff of corruption swirls around road construction projects. Conversations with people about the history of any particular road would always at some point lead to a discussion of theft, embezzlement, nepotism, and shady dealings of one kind or other. And if there was nothing obvious to complain of, there was still the assumption that something underhanded was going on. The roads delivered are never quite the roads that were promised, either in their material quality or in their planned routing. Private interests are commonly assumed to derail the process: politicians and construction companies take a cut; contracts and subcontracts are circulated through friends and families; funds and materials are siphoned off by those who hold positions of responsibility. Those involved in construction processes are assumed to be able to hide such benefits either through technical obfuscations that confirm the quality of an inferior product or by claiming that a partisan ruling on a specific routing that ensures the project will proceed through a certain network of relationships is a disinterested choice. It is not easy to see where the corruption begins and ends. Any kind of involvement in such projects seems to imply some degree of complicity in the endless flow of compromised decisions, bribes, and compensation payments that keep large public works projects going.

Our reflections on corruption take up many of the themes we have explored thus far and perhaps stem most directly from the powerful tension that surrounds all road-building projects: namely, that disjunctive doubling where roads are dreamed of as solutions to a pervasive sense of abandonment and underdevelopment, and yet dreaded for their proven capacity to destroy fragile natural and social environments. People know that road construction projects are risky ventures, and that the material transformations they effect will have unplanned consequences. In many ways this uncertainty contributes to their allure. There is always the possibility that new roads will bring new, brighter, and better futures. We have traced how the charged promise that links connectivity to new flows of wealth can morph into a sense of uncontrolled mobility, environmental instability, and fear at the ease with which others can seize land, resources, and hopes for alternative futures. We have also followed some of the "useful fictions" (Riles 2011) whereby engineering companies seek to create plausible spaces of professional intervention in what are acknowledged as unstable and unpredictable spaces. In this chapter we follow these themes by tracking how attempts to ensure the delivery of public benefit, to control the movement of resources, and to enhance the accountability of public authorities proceed through the use of regulatory instruments. We consider what exactly these instruments achieve in a world where nobody believes they will actually produce a smooth and transparent circulation of knowledge and resources. Our argument is not intended to attribute blame, nor do we assume cynicism on the part of those who apply and administer the regulatory instruments—although such cynicism is certainly in evidence at times.[1] As with our discussion of health and safety regulation, we are more interested in the practices that formal systems provoke, and in the relations they configure between the state, the experts, and the mundane relations of everyday living. As our previous ethnographic description of the enactment of health and safety regulation demonstrated, the construction consortium was caught up in what Marina Welker has described as an attempt to combine a theory of risk with a theory of affect (Welker 2014). In this chapter on corruption and transparency we revisit these tensions and negotiations from a different perspective, looking more specifically at how practices that aim to ensure that roads will achieve progress, growth, and development also generate deep uncertainties, which are manifest in the absolute and shared conviction that public works are always steeped in corruption.

As might be expected, the corruption stories that we gathered in relation to the two central roads in our study are in some ways quite particular. In each instance the stories we were told related to the specificities of that case. Nevertheless, the idioms and practices were also broadly similar. Previous anthropological work on "corruption" suggests that such similarities are to be expected, given that corruption stories mobilize accounts of specific relations to register more general moral and ethical ambiguities. Gupta (1995), for example, was interested in how corruption stories are often clouded by half-truths and uncertainties and provide a language through which people explore conflicting understandings of the practices and moral values of others. Sampson (2005) also suggests that corruption stories signal a more pervasive lack of confidence and sense of insecurity in spaces of regulation and social authority. Expertise enters the picture here. Where experts and regulatory authorities are not trusted, they become the subject of rumor and uncertainty rather than the means by which such uncertainties are allayed. Of course, rumor doesn't imply that something isn't happening, just that it is couched in this nebulous space of uncertainty, where nevertheless an accusation rings true, or the possibility of its truth excites the imagination. In these spaces there is often a sense that "law" and "technical regulation" will not in and of themselves offer a solution—not least because the stories blur the boundaries as Gupta suggests, making it clear that the dividing line between truth and conjecture or legal and illegal practice is highly contingent. This is particularly apparent in worlds where the law is used to enable certain kinds of appropriation, and to exclude and condemn others. From an expert perspective, the problem with rumor is that it is a knowledge form that erodes commitments to evidence and transparency—the possibility of answering back, of democratic argument and scientific experimentation. But rumor is contagious knowledge and spreads complicity in the judgments made and the elaborations entered into as stories circulate and transform.[2] Those who pass on stories participate in the unraveling of established positions and in the process suggest alternatives.

These dynamics resonate, at least partially, with Michael Taussig's concept of the "public secret," those things that are generally known but not openly spoken about (Taussig 1999). Taussig is interested in the relational effects of public secrets, suggesting that they imply a degree of complicity that extends beyond the specifics of a particular case. He argues that the force of the public secret lies in the affective force of the unspoken knowledge that lends tacit support to the judgments and values that generated the need for silence in the first place. We return below to the effects and possibilities of speaking out, and to the ways in which transparency devices and procedures in turn generate their own sidelines and silences.[3] However, at this point it is important to register Taussig's affirmation that all accounts, including the ethnographic, work through the tension between what is revealed and what is left unspoken. Any ethnographic analysis of transparency has in this respect to confront its own interested partiality. We thus need to clarify from the outset that it is not our purpose in this chapter to reveal or investigate specific incidents of corruption on the road projects that we followed most closely. We are, however, ethnographically motivated to try to understand the ubiquity and the energy of corruption stories, their capacity to circulate and to fascinate. We are also interested in the collusions that are both enacted and resisted in the social space of public works.

Etymologically, the term "corruption" carries the connotation of rupture and contamination. Corruption is a disruptive or shattering force, and one that, in social terms, works against a sense of healthy integrity. Corruption is in this respect an important site for anthropological investigation of relational practice. In the context of public works, and in particular of infrastructural projects devised to integrate and create new relations in the public interest (as discussed in chapter 1), the energetic circulation of corruption stories suggests that road construction projects are spaces where particular modes of unease or uncertainty emerge. As we have discussed in previous chapters, road construction projects immerse people—engineers, workers, managers, politicians, local residents—in ambiguous moral spaces: these projects promise to bring progress but threaten to destroy existing life ways; they call for transformation but unleash fears of untoward disturbance; they are generally desired, but there is no core agreement as to what precise social transformation should or could follow from the material changes produced. Much of this uncertainty is gathered into the discussions of corruption and the sense that things do not work out as planned. Corruption, once embedded as an expectation, draws forth technical and regulatory responses that seek to protect the (projected) integrity of the project. Transparency measures and procedures to ensure accountability swirl around construction sites as ubiquitous as the corruption stories they seek to countermand. These measures are focused on making things explicit, articulating expectations and lines of responsibility, stipulating costs and benefits, and ensuring adequate consultation with affected parties. In what follows we suggest that there are key collusions and concealments in this relationship between corruption, regulation, and transparency. Our argument does not in any way suggest a relativist approach to theft and embezzlement. Rather, we are interested in the mutually generative forces of corruption fears and transparency measures, and in the specific ways in which ideas, knowledges, and rumors circulate along the multiple social channels opened by these projects of infrastructural development.

Normative Scandal on the Iquitos-Nauta Road

The Iquitos-Nauta road was famous for its spectacular history of corruption. After the fall of the Fujimori government in 2000, dozens of people were found guilty of illegal activity in connection with the construction of this road, including a previous president of the regional administration, Tomas Gonzales Reategui.[4] However, as we discussed in chapter 1, the history of malpractice stretches back to the original construction project over seventy years ago. Since that time various individuals have used their positions as state officials, within both the government and the army, to profit from the construction process. Political influence and privileged information supported land speculation and ensured routings that would best facilitate the extraction of valuable hardwoods from what had been virgin forest. Indeed, it appears that every time this road was funded, worked on, and abandoned there was something underhanded going on. Some claimed the road should never have been built. Others suggested that indeed it never would have been built if proper procedures had been followed. Most agreed that it was a "political" road. It supported "interests" (*intereses*) and had always been directed at the enrichment of the few at the expense of a precarious and fragile environment. When we first visited the road in 2005, the engineering company contracted to complete the project referred to its work as a repair job. They were expected to put right the mistakes of previous unscrupulous contractors. The legal wrangles over responsibility and outstanding payments continued. Five kilometers had to be left untouched because the responsibility for poor quality surfaces was still being fought over in the courts. At kilometer five, just outside Nauta, a depot of impounded materials was being guarded, but it was rapidly disintegrating in the humid atmosphere of the Peruvian jungle.

We went in search of people who knew more about these stories. We were directed to Jorge, a journalist whose life had more or less been taken over by his determination to bring Reategui to justice. We arranged to meet Jorge in our hotel lobby. The fact that Reategui had been convicted and sent to prison seemed to offer Jorge little solace, and he was clearly still campaigning to convince others of the extent of Reategui's crimes. Jorge was very nervous, constantly looking round, concerned about who the other people coming in and out of the hotel were. They looked like tourists to us. He was anxious to talk to us in great detail about all that he knew, but he was not willing to be recorded. He had brought many documents to show us, but he held back when we asked if we could make copies. Some we could copy, others he was unsure about. He was particularly uncertain about what it would mean for us to have these documents in our possession. What if somebody stopped us in the street? Why would we have these documents on us? Some of them were classified. We could be in trouble if found with them. People in the photocopying shops might not be trustworthy: What might they see, who might they tell? He agreed that we could keep the documents overnight as long as we promised not to take them out of the hotel. He would return for them in the morning. Most of the interview was spent going through the pile of documents as he showed us how to read them. Names were connected up, leads traced for us, and conclusions drawn. The meeting was quite dramatic, and the main message we took from it was that Jorge wanted us to understand that things are not as they appear. Secrets abound, and they are dangerous. This case might appear to be in the past, but Jorge suspected otherwise. Reategui was a powerful politician, and Jorge assumed that he would not stay in prison for long. He thought that when Reategui came out the people who had been involved in his conviction would be in danger of reprisals. We did what we could with the papers that Jorge had given us, trying to piece together the significance of what was being revealed. We came to understand some of the ways in which money can be made from public works. The documents revealed trails of false budgets, false companies, machinery that did not exist or was not moving in the way in which it was supposed to, and works that were accounted for but never actually undertaken. The trail involved a process of connecting names, dates, and figures to build a picture of illicit dealings. However, we could not have found this trail or assessed its significance without Jorge's help. The information was mediated by his contextual knowledge and his interpretation.

Jorge was not the only one who had dedicated himself to struggling against corruption on this particular road. Several people advised us to talk to Pepe, the botanist and campaigning research scientist we introduced in chapter 2. Pepe was not consumed with the anxieties that so preoccupied Jorge, but he was full of passion and anger. Almost as soon as we had shaken hands with him he had his computer open and was downloading PowerPoint presentations to show us, in some considerable detail, what was happening to the incredibly fragile environment through which the road had been built. Although generally despairing about the whole project, he was particularly furious about one section of the route where the road swung right up to the boundary of land that was protected by law as a natural reserve. He explained that this routing was being pushed through by certain politicians who had designs on extracting timber from these protected forest areas. He had tried everything to stop this from happening. His life was consumed with protecting this space, in much the same way that Jorge's life had been taken over by the desire to expose Reategui. Pepe's anger arose from the fact that however many talks he gave, however much information he accumulated and made public—in newspapers, TV and radio appearances, and any other public forum he could get entry to—he could not in the end make the difference that he wanted to. Talking with Pepe we were struck not only by his passion and by how articulate he was about what was at stake for this precious area of the forest (one of the richest areas of insect biodiversity in the world), but also by his sense that things do not, in the end, proceed by rational argument. Information did not speak for itself. Other forces were at work that trumped his clearly exceptional capacities to explore and explain. How to fight "*intereses*" was his dilemma. He wanted to talk with us because he still held to the idea that scientific investigation could win out. If our research was added to his, perhaps it might go further or somehow travel more effectively.

The dedicated fervor of both Jorge and Pepe was in many ways quite exceptional. Most people we talked to were more resigned to the ubiquity of corruption than they were. Nevertheless, their responses were quite typical in other ways. First, there was common agreement with the basic premise that corrupt practices lurk behind façades of respectability. Nothing is as it seems, and there are always revelations to be made. This tension between the hidden and the revealed is a daily drama rehearsed in the Peruvian media.

Corruption as scandal sells newspapers. Suspicion and expectation of wrong-doing are commodified, and the mass production and circulation of corruption stories in many ways render them quite mundane and unremarkable. There is a general assumption that large projects that work through layers of subcontracting constitute what we came to think of as a "normative scandal," the habitual expression of the sense that there are always private "interests" behind public works. Those in positions of power and authority, particularly engineers and politicians, are deemed most likely to take advantage of the connections and relationships that infrastructure projects establish. Knowledge of a prospective route might, for example, allow the identification of a lucrative site to buy land or an opportune moment to sell. A well-placed word or gift to a contracting engineer might help a friend or family member get work or secure a "deal" to persuade another of the viability of granting a contract. The sourcing of materials or the siting of a camp also present choices and require judgments that rest on an opaque expertise. Expert knowledge is by definition not easily accessible; it is not transparent or directly apprehensible to the untrained or uninitiated. It is set apart in its language, its metrics, and its guiding principles, and it creates a space of dependency and mutual distrust.

Both Pepe and Jorge were also concerned, however, that making things explicit and accessible does not necessarily make a difference. They were both working from the assumption that there were hidden dealings and relations that should be exposed so that others can see for themselves the private interests at work. However, experience had taught them that the evidence they compiled was insufficient. In both these cases knowing what they did was not enough to change things. The evidence incriminates, but it does not resolve the issue. The problem is that what they reveal is at some level not surprising to most people. The scandalous practices are depressingly commonplace, the "secret" dealing already anticipated. The specific information that they work to make public simply confirms the rumors and half-truths already in circulation. The scandals about specific individuals do not displace the deeper and more pervasive sense that corruption is at another level just a way of talking about how things actually get done in practice.

So how might Taussig's notion of the "public secret" help us analyze these relations of distrust and revelation? Taussig's work centers on a rereading of Julian Pitt-Rivers's *People of the Sierra*, which describes life in a village in Andalucia in Spain during the Franco era. Taussig is interested in what is and is not revealed in that classic ethnography. Pitt-Rivers chose to write about masculinity, kinship, and *compadrazgo*, and his work paid attention to how people managed their relations while living under the strain of an oppressive state. What he did not discuss was the well-known fact that this town was at the center of the anarchist movement in Franco's Spain. These political commitments were an open secret, never discussed by local people or by their ethnographer. Taussig argues that the public secret of the anarchist leanings of this small town had a collusive quality that served to "lubricate the workings of the state." The common knowledge was never spoken, but it always could be, and the threatening potential of that knowledge was thus amplified. People lived in the knowledge that state agents knew but did not act on the knowledge, or did not act yet. Knowing that others know but do not act (or even acknowledge that they know) reproduced and magnified the sense of insecurity. In these circumstances a degree of protection or public safety is afforded by turning a blind eye, by knowing what not to know (Taussig 1999), but such protection has the effect of enhancing the sense of danger. The "public secret" in Taussig's terms is about a concealed relation.[5] The concealment is not, however, primarily about the information that an antagonist could reveal to destroy another, but refers more potently to the concealed collusion that creates a space of uncertainty and compromise in which all are somehow implicated. Together, oppressor and oppressed produce and reproduce the sense that there was something unspeakable going on in that town.

The scenario is certainly familiar to those who knew Peru in the 1980s and 1990s, when the war between Shining Path and the state was at its height. These were times when duplicity was deployed as a weapon by both sides. The army would appear in Andean villages dressed as guerrilla fighters to test loyalties, and the insurgents would do the same, dressing in army uniforms to weed out collaborators. Local tensions were played out through accusations of terrorism. There were betrayals and tip-offs. Many people were murdered, many disappeared, and although many suspected, and some knew for sure, there was no general discussion of who was involved and how. After the war the Truth and Reconciliation Commission aimed to fill these silences in an attempt to counteract the terror of uncertainty by making things explicit. However, for many people this was not the way forward.[6] Some acknowledged that it was better not to know. It would be too difficult to go on living next to the person who had betrayed your kin, too hard to bear the sure knowledge that your kin had betrayed one another. For many it was preferable not to know for sure, even when there were suspicions. The sensitivity experienced in Andean villages in the aftermath of the war stands in stark contrast to the current determination to conduct politics through the language of corruption accusations. Political opponents routinely accuse each other of corruption. In many ways corruption accusations have displaced other modes of political argument. The government now sells official kits that citizens can purchase and use to instigate proceedings to oust their elected authorities. In this charged atmosphere of political distrust, the concept of the public secret might appear to have been superseded. There is little that is secret; it is all about openness and publicity. And yet while the public denunciations have become utterly familiar, there is a growing sense that such excessive publicity in turn conceals other more significant relations. These, we would argue, are the very relations that produce the conditions that foster both the sense of ubiquitous corruption and the commodification of scandal revealed, and they are at work in and through the regulatory instruments that are applied to public works projects. Indeed, as Kregg Hetherington points out, the perpetual need for further "unmasking," which the combined forces of corruption accusation and transparency responses produce, skates over the fact that "corruption" and "transparency" are very different kinds of things. "Corruption is an ethical lapse, while transparency is a visual quality that an object is said to have if one can see beyond it to something of greater interest" (Hetherington 2011, 153). Transparency practices thus produce, or make evident, the very thing that they are called on to measure, reproducing perceptions of corruption in "self-referential political loops."

With this in mind we would suggest that anthropological accounts of corruption could take one step further Akhil Gupta's (1990) idea that corruption stories register moral ambiguities and uncertainties. The explosion of public interest in corruption is happening at a particular historical juncture that bears thinking about. Deborah Poole has suggested that corruption stories do not simply register moral ambiguities and uncertainties; they register the quite specific moral and political problems that arise in relation to neoliberal values—particularly the ways in which notions of the "public good," and the related understandings of "locality," "community," and "sovereignty," are reconceived and mobilized by the state in a country such as Peru (Poole 2005). Indeed, as she points out, corruption stories multiply in particular kinds of spaces, spaces that set themselves up as offering a clear division between public office and private life, but which simultaneously present the entanglements of public and private benefit as necessary and as providing the best form of public value. It is perhaps unsurprising that we find contemporary infrastructure projects particularly subject to corruption allegations. This is not to suggest that corruption practices are new—documentation from the earliest colonial times talks of those who sought to make profit from that which pertained to the Crown, and subsequently to the republican state. It is the ubiquitous deployment of corruption accusations, and the understanding that a generalized corruption permeates all levels of society, that we are interested in exploring here.

The corruption stories that surround the Iquitos-Nauta road all came to a head when Fujimori left Peru in 2000, taking huge sums of money with him and leaving his closest adviser and head of the Peruvian intelligence service, Vladimir Montesinos, in jail after one of the most publicized corruption trials of Peruvian history. Gonzales Reategui was closely associated with the Fujimori regime. Fujimori had placed him as president of the CTAR-Loreto.[7] While the road construction project progressed in this period, it was also the period when critics, such as Jorge, were drawing attention to misconduct, false contracts, and the sequestering of funds. Gonzales Reategui also headed the regional authority in controversial peace negotiations with Ecuador. Peru's agreement to cede territory to Ecuador, which was ratified in the 1998 peace accord, was directly related to the release of funds for the construction of the Iquitos-Nauta road, and, some argue, to the direct enrichment of many local officials. Gonzales Reategui had by now moved to Lima, where he was heading the Ministry of the Presidency, the office of state charged with the establishment of norms for the regulation and coordination of decentralized government. When Fujimori left office Gonzales Reategui was brought to trial in Iquitos and convicted on corruption charges in association with the construction of the Iquitos-Nauta road.

In 2005 Fujimori returned to South America in a remarkable moment of political misjudgment. He was imprisoned in Chile and subsequently extradited to Peru, where he is currently serving a twenty-five-year prison sentence for human rights violations, corruption, and embezzlement. However, his reputation for delivering public works was also openly and actively used by his daughter Keiko in her 2011 presidential campaign. And perhaps even more remarkably, she almost won, coming a close second with over 48 percent of the vote. Her partisans are supporters of the Fujimori brand and declare themselves Fujimoristas. Despite Alberto Fujimori's appalling record of human rights abuse in the war years, and the deep corruption that took root in the ten years of his authoritarian regime, many people remember the measures he took in his first term of office. He quadrupled the minimum wage, built many schools and roads, and instigated neoliberal reforms, which for some are associated with the opening of new markets and new opportunities. Many people remember that under his leadership the economy was strong, people had employment, and the state invested in basic infrastructures. The mantra that "he might have stolen, but at least he did things" is a familiar one that is frequently used to explain continuing support for corrupt politicians.[8]

In 2011 Keiko Fujimori's opponent, Ollanta Humala, promised to attack corruption. The promise was a popular one, but it also was an unnerving one that was linked by the right-wing media to a growing fear that Humala might also bring a mix of military-style authoritarianism and populist nationalism that might scare capital away. It was hard to establish to what extent Fujimori's authoritarianism and the human rights abuses and corruption within his regime worked against him. There were other things in his favor. His disregard for human rights had coincided with the capture of the leader of the Shining Path (1992) and the release of all but one of the hostages who were held for three months in the Japanese Embassy by fighters of the Túpac Amaru Revolutionary Movement (1996–97). In Keiko Fujimori's 2011 campaign the right-wing press adopted a "better the devil you know" attitude. A fundamental lack of trust in either candidate fed these seeming contradictions. What can you know of these candidates, and what might they do once in power? What do the campaigns and the public debates reveal? How can the notion of public information have credibility when the media is so blatantly partisan? The election campaigns worked through paths of rumor and scandal that were every bit as plausible as other, more official kinds of information. The key questions remained unanswered, and were ultimately unanswerable: Which of them would provide security, welfare, and income, and at what cost?

The intractability of the problem is exacerbated in the contemporary world by a requirement that all governments face, of negotiating a national advantage by brokering relations with more powerful international interests. This deeply compromised space of the national economy directly impinges on perceptions of corruption and on the measures that are put in place to combat corruption. International investors are concerned to ensure the safety of their investments, and a swath of regulatory measures accompanies all such investments, particularly those made to public bodies by large international banks. As public works, road construction projects are a prime example of such investments; thus, despite the fact that such projects appear so compromised by corrupt practice, they are also tightly wrapped in anticorruption regulations. It is to this ambiguous relation that we now turn.

Anticorruption Regulation

The corruption scandals that emerged in the 1990s in Peru coincided, on a global scale, with the founding of Transparency International, an organization set up in 1993 by a retired World Bank official, Peter Elgen. Transparency International's account of its own history starts with the observation that "in the 1990s, corruption was a taboo topic. Many companies regularly wrote off bribes as business expenses in their tax filings, the graft of some long-standing heads of state was legendary, and many international agencies were resigned to the fact that corruption would sap funding from many development projects around the world."[9] Responding to these conditions, transparency was to become the new public virtue of post–Cold War geopolitics, juxtaposing the centrality of openness to democratic systems with the secretive workings of totalitarian regimes. Transparency International now has representation in over one hundred countries to tackle what is recognized as a global problem, one that Sampson identifies as an era of globalized corruption and globalized anticorruption, where everyone is corrupt and everyone is against corruption (Sampson 2005). Sampson asks a very pertinent question: "Why, in the midst of no holds barred neoliberal efficiency and organizational downsizing, do we also find more institutions consciously trying to 'do the right thing'?" Sampson is interested in how anticorruption is adopted both as a moral project, with the aim of making the world a better place, and as a rational measure, with the aim of making the world a more efficient place. Organizations such as Transparency International work to make the costs of corruption explicit. There are political costs (corruption is a major obstacle to democracy); economic costs (corruption depletes national wealth and hinders fair market structures); social costs (corruption undermines people's trust in political systems); and environmental costs (the relentless exploitation of natural resources has far-reaching, potentially irreversible consequences).

Public officeholders in Peru are highly aware of these discourses. The folding of such calculations into an increasingly financialized mode of government supports the ways in which corruption accusations have emerged as a key political tactic in recent years. The destructive consequences of corrupt practices are increasingly invoked to explain all manner of political, economic, social, and environmental problems. Against this background of a globalizing mission to eradicate corruption, governments across the world have begun to institute specific measures to control public institutions—most particularly to control the ways in which public money is spent. In Peru one such instrument—the one that most directly relates to infrastructure projects—is the Sistema Nacional de Inversión Pública (SNIP, the National System of Public Investment).

The SNIP is an instrument of the devolved/neoliberal state. It has been developed in Peru with explicit connection to political processes of decentralization—specifically, the transfer of responsibilities and financial resources to subnational governments. It is a tool that has been developed in direct relation to institutions such as the World Bank to support the loans that are increasingly administered by local instances of the state. Housed in the national Ministry of Economy and Finance, the SNIP is applied to all public works—whether commissioned by the central state, any of the twenty-five regional governments, or the 1,650 local municipalities. The SNIP is explicitly designed to orient the transformational process of development. On the websites and guidance documents issued to local government officials the core aims are clearly stated: (1) the promotion of efficiency in relation to the use of investment resources; (2) the building of a sustainable knowledge base for the management of public investments; and (3) the maximization of the positive socioeconomic impact of public investment. The SNIP also has legal force. It requires the investor (the state and those they contract with in public-private partnerships) to make explicit the components of a project, to fully rationalize the investment, and to systematically display evidence of strategic planning and consensus-based decision making. In this respect the SNIP is quite clearly a governance device. It restricts the ways in which projects can be imagined and justified, and its legal force renders local officials vulnerable in the spending of public money, as mistakes are construed as infractions of the law.

As a knowledge space and a participatory device the SNIP exemplifies the features that Green (2010) has identified as characteristic of participatory methods, whereby the social body is configured as both "agent of transformation and object of intervention." In other words, these devices are integral to processes of decentralization, whereby the responsibility for public works is devolved from central state agencies. Responsibility comes to rest with local citizens and their elected representatives, who are required, through the demands of these instruments, to become actively involved in changing the conditions of their own lives. At the same time, the central government uses the SNIP to maintain a tight control over public finances. The SNIP in essence is the public document through which expenditure on public works is made transparent. Projects have to be presented as responses to problems identified by specific populations, beneficiaries have to be named and counted, and the competences, technical capacities, experiences, and resources of contractors and subcontractors have to be made explicit.

Given the experience on a project such as the Iquitos-Nauta road, the instrument is welcomed by the many public officials and engineers in Peru who are charged with delivering public works at a time when awareness of corruption renders all open to suspicion. Securing a SNIP number for a project is essential (although not sufficient, as we explore below) to securing social validation, and it is generally seen as a way of attracting investment. The system is designed to eradicate nepotism and the arbitrary awarding of contracts for projects that do not have general public support. However, the SNIP is no guarantor of transparency. Strathern (2000a) refers to the "tyranny of transparency" in her analysis of the ways in which any attempt to render relations explicit will always also draw attention to all that is not yet revealed: "what is invisible is what is simply not yet made visible." There is always more to be revealed and "further realities to uncover." Making some things explicit only displaces, rather than erases, what is implicit.[10] When it comes to road construction projects the SNIP begins to draw attention to the intrinsic multiplicities and ambiguities that construction projects entail. For example, the electronic data fields of the SNIP require that "the social problem" be articulated as both singular and stable. There is no space for complex aims or for the awareness of the contingencies (material and social) that emerge in the process of construction. The regulatory requirement for clarity and transparency thus has unsettling effects. Once a project is deemed feasible, via the application of an instrument such as the SNIP, both the problem and the solution have been defined into existence in ways that leave little room for maneuver. Budgets are fixed, and the potential for adapting to circumstances that unfold in the course of the project—something that we have shown to be central to road construction projects—is limited by a lack of resources for such contingencies. For the SNIP to operate successfully, everything needs to be worked out in advance.

The information collected within the SNIP draws on an evidence-based paradigm, and yet road construction is self-evidently carried out in a field of complex and overlapping knowledge forms. Green (2010) points out how the participatory methods deployed in instruments such as the SNIP are explicit attempts to elicit situated knowledges. However, the goal of using the studies as a basis for intervention creates a situation in which situated knowledge (which acknowledges the specificity of the particular relational dynamics through which information becomes significant) is made subservient to evidence-based knowledge (which utilizes objective measures to produce a picture of the external world into which the intervention will be made). The SNIP does address this opposition, but it does so only in the sense that it requires the transformation of situated knowledge into statistical projections. There are explicit formulae for generating credible projections of production, population, or any other relevant circumstance, but they all must be converted into data that can be considered in terms of trend, direction, and change. Significant relations are inevitably left out of project specifications. It is one thing to specify the entities involved (the organizations, settlements, land holdings, watercourses, and so forth) and the geographical areas concerned (although these are often quite arbitrary political divisions). But the SNIP also requires some elaboration of the antecedents to "the problem," accounts of prior attempts to solve said "problem," and, most problematic of all, why it is in the interests of the community to solve the problem and why public expenditure is justified. These are highly political questions that are folded into the neat diagrammatic format of the SNIP as if all the data and the figures gathered are of a kind, and thus commensurate. What is concealed in these transparent documents are the complex political negotiations and the lost voices from which the rational and singular statement of intent is produced.

In this way the SNIP harbors huge potential to generate uncertainty:

It may condense long processes of public consultation and erase dissenting voices, it may appear without sufficient consultation such that dissent was never voiced, or it may articulate a problem that is generally recognized, but for which any given solution is likely to be deeply controversial. The process requires cost/benefit analyses but the cost-benefit metrics of alternative solutions are highly political and replete with value judgements obscured by the apparent self-evidence of the metrics. Suspicions of a lack of transparency in this crucial field of public decision making thus multiply because the figures themselves make evident all that is not seen, and not recorded in the official paperwork. The sense of suspicion is further enhanced by the tension between insistence on the importance of fulsome public participation in the process, and the legal requirement that such participation is subject to the technical framings (whereby figures, costs, and relative worth are assigned) of expert committees. (Harvey, Reeves, and Ruppert 2013)

In the following section we describe a meeting in which some of these complexities were played out in the context of the construction of the Interoceanic Highway.

Protection and Responsibility—after the Event

Hannah was on a plane heading to Puerto Maldonado for a meeting between the construction company and various environmental NGOs, and found herself sitting next to Manolo, a representative of the World Wildlife Fund (WWF) who coincidentally was on his way to the same meeting. They got to talking and he drew a diagram to explain how he saw the field in which they were operating (see figure 11).

Manolo explained how the government, the international financial institutions (IFIs), and the engineering companies all work together on major road construction projects. Conservation NGOs such as the WWF tend to be brought into such projects through the IFIs. Many banks, including Credit Suisse, which was directly involved in this case, have a code of practice that any project they invest in must adhere to. The finance arrangements for the Interoceanic Highway had been worked out as a public-private finance initiative. Not all the money required was in place when the project started, and as different monies came on stream at different times, so the conditions of the funding arrangements varied over time.[11] Odebrecht, the lead partner in the CONIRSA consortium, had secured a $200 million loan from the Andean Development Corporation (CAF) to get the project started. In 2006 it borrowed money from Credit Suisse to meet its repayment obligations to CAF, and at this point the WWF was drawn in. Credit Suisse contacted the

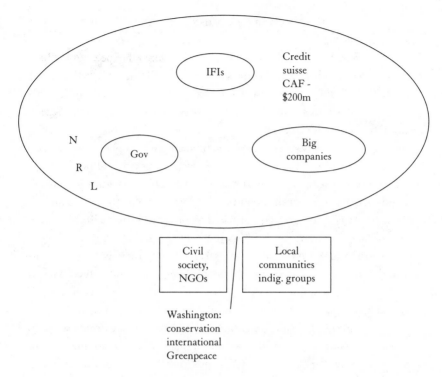

Figure 11. A diagrammatic representation of the key institutional actors involved in infrastructure funding

WWF, asking them to produce a list of social and environmental impact criteria that CONIRSA would have to meet to be eligible for the loan. This was proving problematic, and Manolo was on his way to Puerto Maldonado to continue the negotiations. He explained that on the whole the large companies like to keep dealings with NGOs and with local communities quite separate. The NGOs advise the companies on how to meet their international obligations, and the parameters are relatively clear and well defined. Engagement with local communities is far more complicated. Local expectations of the company often exceed what the company is able or willing to provide (as we outline in more detail in chapter 6), and there is ambiguity at every stage about the overlapping responsibilities with the various levels of the state—the national, regional, and local governments. However, as the meeting to which Hannah and Manolo were headed reveals, engagement with the NGOs and the banks also gives rise to confusion and misunderstanding.

Critics of the Interoceanic Highway project deplored the speed with which the arrangements between the government, the IFIs, and CONIRSA had been settled. President Alejandro Toledo had rushed the project through parliament, anxious to secure this major public work as a personal legacy before the end of his term of office in 2006. Not all the necessary studies had been completed in advance. Indeed, the financing was in place and the contracts signed before the environmental impact assessment had been carried out.[12] The idea was to carry out the environmental impact study piecemeal, keeping ahead of the construction process, but inevitably the road could no longer be considered as a single space. The piecemeal study also precluded the possibility of taking into account more widespread and longer term environmental and social impacts. Most serious among these was the proximity of the road to the Tambopata National Reserve, which the road will inevitably transform by making the reserve more accessible. Large trunk roads invariably provoke the construction of smaller side roads, which are projects that local communities can fund and construct with little external support. Manolo was not that sure of the overall rationale for this road. He was not convinced by the story that the Brazilian government wanted an export route, as analysts had argued that the Panama Canal would still offer a more cost-efficient route to China. He wondered whether it was not in fact Odebrecht or IIRSA that had the most to gain from the project. The road would certainly further IIRSA's interests in opening up transcontinental telecommunications and power lines, and Odebrecht was also interested in the hydroelectric opportunities the region offered them, as we discuss in the conclusion.

The meeting in Puerto Maldonado was chaired by Roberto, head of the environmental division of the CONIRSA consortium. He discussed the second phase of the environmental impact study, which was getting under way. He was looking to the NGOs represented at the meeting to be actively involved in this process. As the meeting proceeded the issue of participation took on increasing salience. It mattered to the company who was there, who they talked to, and who they could account for as groups that had been consulted. In this respect, local community groups were folded into the process. Those groups that coalesced around specific economic development initiatives were most welcome, as these were the communities to whom the road would be most likely to deliver a direct and positive impact. They could and would serve as exemplars of the benefits.

There was much talk of social responsibility. The WWF were pushing CONIRSA to outline the measures it would take to minimize the indirect impact of the road, while ensuring development for local communities. The CONIRSA employees were in a difficult position. Gradually they began to talk about something that everybody around the table already knew but had not mentioned until that point. There had been a slight irregularity with this project. The SNIP procedures had been somewhat abridged. Only the profile and the prefeasibility studies had been completed. No final engineering plan had ever been elaborated. This fact carried a further unspoken implication—namely, that CONIRSA had been awarded the contract before the specification had been agreed. These issues had to be aired because CONIRSA wanted to stress that it had had to produce the final specification after the event. The conversation clarified for us why there had been so little construction going on when we first began our study of the Interoceanic Highway. CONIRSA was catching up. The environmental impact study was one of the pieces of this final specification. It had been subcontracted to another company after the main construction contract had already been awarded—and it too had been done in a hurry.

From CONIRSA's point of view, it was making the best of things. There was no discussion of how it had secured this lucrative contract. All discussions started from the point at which the contract was awarded. This point was the fait accompli that all now had to work with. The international financial institutions and their associated environmental NGOs wanted a fuller account of the presence of indigenous communities, existing and projected population movements, and existing and projected economic activities, but CONIRSA did not have the information to provide on these specific points. However, there was still a lot of information—in some ways too much information, as it turned out.

In the course of the meeting it was revealed that the subcontractors had produced two environmental impact studies, and there were also other independent studies in existence. For example, the Ministry of Transport had spent two years compiling a feasibility study that had a substantial chapter on the environment. The Andean Development Corporation had done its own study, but the CONIRSA personnel had not seen a copy. Somebody suggested that it would be a good idea to try to get all these documents together and to work out precisely what was and was not known about the space in which they were working. At this point the picture began to get more complicated. Roberto pointed out that the contract for the construction involves many different parties. There are separate organizations building the bridges, and not all the bridges are contracted to the same company. Each one will have its own loans, contracts, and obligations, and their impact studies will be separate. CONIRSA did not have responsibility for all the construction work. Furthermore, although it might be assumed that the government would take on a supervisory role that would oversee the work in general, the timeline of the project suggested that there would be many changes at the political level before the road was completed. Without either a ministry of planning or a ministry of environment at this point (the latter was subsequently founded in 2008), such continuities could not be provided by the central state.

The meeting returned to the theme of participation. CONIRSA acknowledged that there had been far too little consultation with NGOs in the first environmental study, but it hoped that now in this second phase they could share information and listen to all the different points of view. Dialogue became the focus, the need to talk to as many organizations, committees, and representatives as possible along the road. Someone asked how exactly the NGOs and representatives of civil society were going to contribute if the studies for the second phase were already under way. CONIRSA acknowledged that it was a little late to rethink the basic shape of the project, but the consortium still needed input on the management of the road for which they have a contract for a further twenty-five years for both toll income and maintenance responsibility. The agenda had shifted subtly. The issue was no longer environmental and social impact, but how local people were going to live with the road, how they would benefit from it, and how they could be encouraged to use it responsibly. In these discussions the road had become a passive and neutral space. Responsibility for the success of the road now lay with local communities, and the responsibility of the construction companies, the government, and the IFIs faded. Their agency was highlighted with respect to the financing and the execution of the design. They would deliver the road. The finished road thus became the space where something could and should be done. However, in this framing it was now the community that was responsible. The clear problem that local communities had not been included in any of the prior discussions—even though they had campaigned vigorously for a road—was again not up for discussion. The question that concerned the meeting was what information should be made available to local communities. What did they need to know? Who, out of the many different national and international organizations involved, should they be in conversation with? It was agreed that an integrated vision of the road was needed.

This notion of integration, which we visited initially in chapter 1 and which has recurred as a theme throughout this book, emerges at moments when awareness of difference and disjuncture comes clearly into view. Integration is the idiom through which to address the social problem of fracture, difference, and multiplicity. In this meeting, it was the excess of documentation that provoked the call for an integrated vision.[13]

As the meeting progressed the excessive numbers of studies became a more tangible problem than the inadequacy of any particular study. There were simply too many studies in circulation. Some were openly available and known about; others belonged to specific organizations. Nobody seemed to have access to everything, and most people did not even know what documents existed. There was a strong sense of distributed responsibility and fragmentary structure. The production of more studies did not promise to make things clearer or to get the project back on track. However, the discussion about the documents did reveal that there was no single axis of responsibility and no agreement on how to locate or track the benefits of this project. Who would decide where or how to locate the public good? How might the "public good" contrast to the sense of "private interests" in a world where public-private arrangements are seen as the "best" solution to infrastructural challenges, and where the enfolding of private interest into public projects is heralded as the way forward? The problem is not exactly a refusal to consult or to take participation seriously. It seems rather that the space of participation is already crowded with too many different interests.

The Regulation of Multiplicity—Corruption and the Public Secret

If we look again at the diagram that Manolo drew for Hannah in the plane, we can see the following configuration of overlapping but dissimilar interests:

- The government wants political effect as fast as possible but also has to consider changing social and environmental impacts.
- The banks want a viable business proposition, which might include environmental factors that are important to the sustainability of their ethical investment funds. But there are various banks, entangled in the funding of each other's loans, and the terms of the financing deals change over time as new loans are taken out to service previous ones.
- CONIRSA also needs a viable business proposition and needs to meet the regulatory demands of the banks and the government, and its business propositions have to accommodate the interests of its client (the government) and its funders (the banks).
- The NGOs are an extremely diverse grouping that serve government, banks, and construction companies in different ways, but also include campaigning groups totally opposed to the construction project or intent on securing benefits for local communities, or both.
- Civil society is also not an integrated entity but encompasses all citizens and their elected representatives across the political spectrum. There is no single answer to what it is that civil society wants or expects from the road. Political leaders represent their constituencies with recourse to an (imagined) majority opinion.
- Alongside all these groups there are many other people—less obviously a part of civil society because they are less organized, but nevertheless present and looking to make a living from the roads. These people swiftly come to constitute the problem that other agencies focus on. They appear as tangible barriers to progress—who could nevertheless be made to act otherwise. These people became the focus of conversation. For example, the meeting considered how Brazil nut growers on the Interoceanic Highway are likely to react to the in-migration that the road will produce. Currently the majority of these nut producers live on land that was given in concession for Brazil nut production but not for agriculture. They are, however, likely to be tempted to sell land (which is not theirs to sell), and retrospectively, if challenged, they might claim that the land was invaded by the incomers. Loggers are another and different problem. The loggers are an economically powerful group, able to build side roads and run semi-clandestine extractive operations through the payment of bribes. The meeting participants worried that state authorities might not have the agility to respond quickly enough to these threats.

The Interoceanic Highway was never subjected to the regulatory procedures of the SNIP, and thus it did not have to find a way to draw together these diverse interests into the articulation of a singular and common project. The lack of regulatory control has proved to be a problem, but it has also afforded us the ethnographic opportunity to observe the difficulties that any such procedures would inevitably have encountered. A SNIP could have been produced, but such a document would have required patching up and glossing over the gaps and discontinuities that clearly came into view in the course of the Interoceanic construction process. The case of the environmental impact study is important. What we found here was that there were multiple studies, but they did not cohere, and they could not be made to add up to a single view. They were not all accessible, and they were carried out by different companies in relation to different briefs. In all, we counted six studies in circulation. The lack of common terms of reference for these construction projects is not surprising: they have emerged over many years, through many different political regimes, and enfold many different interests and expectations. The desire for such common ground is similar to Jürgen Habermas's utopian promotion of the notion of "common language" as the basis for an effective public sphere, a space in which all opinions can be voiced and evaluated openly and in dialogue. The problem with Habermas's utopian vision is that it requires a language without history, nuance, double meaning, or a capacity for implied insult or ambiguity. Nevertheless, the idea is one that many have shared over the years, and it is this notion of forging a common language and a common purpose that inspired Richard, CONIRSA's communications manager.

Richard insisted to us in an interview that the key to a successful development project of the kind they were engaged in is the constant need to improve communication by encouraging continual feedback. He spent his time visiting communities up and down the seven-hundred-kilometer stretch of road and talking with all the other interested parties: the banks, the government representatives, and the NGOs. He described himself as a great advocate of transparency, and explained that he wanted people to know that CONIRSA has nothing to hide. His words might well be genuine, but the problem is that from a local perspective they come too late. In the communities along the route people know that power works in nontransparent ways. They do not necessarily seek to change that. Their demands are not primarily for out-and-out transparency. What they do want is some tangible benefit from this project now that is going ahead anyway, and they are looking for some gesture of involvement. Minimally, they look for work and for improved living conditions. However, as we will discuss in chapter 6, even these demands can be deemed unreasonable, rapidly converting a critical public into an impossible public that is not understood or even recognized as an appropriate interlocutor.

Infrastructure projects make particularly clear the difficulty that neoliberal states have in delivering on a social contract that works from the basis of inclusion. Inclusion and generalized participation require greater intervention by the state to control the private interests of capital investors, and yet the ambition of a neoliberal state is to release capital from undue regulation in order to encourage "growth." In this context the public component of public works take a strange turn. Excluded from the original terms of the financial arrangements between banks, construction companies, and central state agencies, the NGOs, community groups, and local government agencies nevertheless become responsible for the effects of projects set up in terms that do not return the benefits they were expecting. And they are very unlikely to ever do so, for these public benefits were never clearly specified. Such benefits should ideally be worked out through the regulatory procedures, such as the SNIP. However, as we have seen, the SNIP itself requires a coverup, an abstraction away from the complexity of interests and expectations in order to produce a singular aim and an agreed solution. This failure to register and accept difference once again sets the ground for further corruption accusations. For those who perceive no clear public benefit, questions arise as to who did benefit. If the road does not deliver for them, who is it delivering for? The answer is known in advance. The road must be serving private interests. In the case of the Interoceanica the narrative of Odebrecht's gain became consolidated over the course of the project. Odebrecht was the partner in CONIRSA that seemed to onlookers to emerge as the winner. They had the contract (they were paid for the work) and they had a twenty-five-year investment opportunity. The only thing that now stands in their way is the possibility of local protest. It is little wonder, then, that they are concerned with openness and feedback and with propagating the idea that the public interest can be accommodated in relation to the interests of their company— as long as other (competing) private interests, such as the nut growers and the loggers, do not get in the way and disrupt things.

With this tension lying at the heart of contemporary infrastructure projects, we might ask whether instruments such as the SNIP can ever be expected to control *intereses*. Expert knowledge might be expected to limit the multiplication of what could be known. The authority of the expert imposes closure in this respect. However, the participatory dimensions of the SNIP blow this possibility out of the water, for this procedure explicitly requires that expert knowledge be introduced to a relational space in which everybody already knows that they do not trust the experts. The viral stories of corrupt engineers and politicians already have purchase. Credibility rests on prior knowledge or trustworthy recommendations before it rests on expertise. Thus, when the communications manager tries to tell people that what really matters is that they are all in good, regular communication, he totally fails to address the things that will consistently undermine this possibility. These are not primarily the private interests of the self-serving individual, but the quite specific different social interests that each participant brings to the table. The fact that the SNIP can make such divergent interests commensurable by rendering them all as data appears to feed the sense that corruption—or at least private interest—is the main problem.

This observation brings us back to the notion of the public secret and the question of the complicities and the obfuscation that corruption scandals sustain. On the whole, discussion of these scandals focuses on personal failings, on the corrupt individual who destroys or shatters the collective social project. What is much less often discussed is the approval that these large public works initiatives elicit for something that is always only a statistical projection. In the process of manufacturing a sense of common desire, interests that are only ever superficially compatible are drawn together through the language of collaboration and participation. This is the social space in which the SNIP seeks to operate—but it is also a social space that exists only in the imaginary of an instrument such as the SNIP. Once a project is under way, the singularity of the SNIP is undermined: too much specificity emerges both in the relations of participation and in the constant generation of information that is never "clean" but always emerges from specific conditions, contracts, and understandings. These relations appear corrupt when the roads fail to deliver on their utopian promise of meeting everybody's needs. Our understanding of the corruption that plagues public works is that the public secret of incompatible interests works to amplify the sense of corruption and exasperates all those who try to iron things out by improving regulatory procedures. Or, to put it another way, politics is the public secret of technical solutions. In a world where people have lost all trust in politics, technical regulatory systems are brought in to stave off the overtly political negotiation of difference. However, the social effect of this move is to debilitate the possibilities of forging new social entities that acknowledge and work through difference, and in this sense these technical measures render politics unspeakable. At the same time, transparency measures that encourage the production of documentation to display good practice, or to manifest evidence of corruption, also always come up against the problem that the information mobilized in this way never simply maps a stable or singular external reality. Further documentation, interpretation, and contextual knowledge has to be brought to bear, and the political reemerges as internal to those systems that seek to establish their credibility by its expulsion.

PART III

The Modern State

Promise and Deferral

Chapter 6

Impossible Publics

Quince Mil is not the kind of place that guidebooks encourage their readers to visit. From the perspective of a traveling tourist there is nothing much to see or do. Its main claim to fame is its self-image as the town with the highest rainfall in South America. The scene on our arrival partially confirmed the prior impression that we had garnered from our various travel guides. Although it wasn't raining, it was hot and humid, and the cumulonimbus clouds were swelling threateningly overhead. At midday the town seemed quiet and empty. Most of the shabby buildings were spread along a single unpaved street through which all traffic from the Andes to Puerto Maldonado has to pass, although traffic on this stretch of road in the middle of this day in late March was intermittent. Nevertheless, the street was singularly oriented to servicing the little passing trade there was, with bars and restaurants, repair shops, a police post, a couple of hotels, and a gas station strung out along the muddy mile-long strip.

Quince Mil is on the route of the Interoceanic Highway. It lies at about 950 meters above sea level, thus marking a midpoint between the Andean section of the road's route and the Amazonian lowlands into which it leads. We first visited Quince Mil during a journey we were taking along the entire route of the Interoceanic Highway construction project. We had only scheduled a day in Quince Mil, so as soon as we arrived we began to wander around to find people to talk to and soon fell into conversation with a group of men gathered by the town hall. They responded to our eager queries about where to find local people who knew about the history of the road by disavowing us of the assumption about the presence of local people in Quince Mil. There were people who had lived there a long time, but there were no locals in the sense of people who were originally from that place. When we ventured, in response, that in any case it might be useful to talk to the mayor, we discovered that she was a particularly problematic outsider.

The mayor, it transpired, had not even been voted into office. The previous mayor had installed her as his deputy just under three years ago. When he left office, in what appeared to have been a botched referendum in which a confused electorate got rid of their chosen representative by mistake, she

Figure 12. The Interoceanic Highway at Quince Mil, 2006

had taken over the job. In the opinion of the two men who we had engaged in conversation, this was a disaster: she did not operate in the interest of local people, and she did not have a local mandate.

As dusk fell we discovered, with the waning light, that there was to be no electricity that night. It was common in many of the towns along the route for the electricity, or *la luz* (the light), to be cut off during the day, when artificial illumination was not needed, but we were told that this was the fourth consecutive night that the town had been without power, and there was no clear indication as to when the supply would be restored. As it got darker, impromptu congregations began to form in the street, with groups of people milling around the open doors of the few cafés whose generators were chugging out an unsteady current, which was transformed into a weak florescent glow from the low-energy strip lights. Rumor had it that the mayor was preparing to leave for Cusco that night to arrange for a mechanic to come and fix the broken motor in the generator. A couple of hundred kilometers up the road at San Gaban was a huge hydroelectric plant, which harnessed the water that region had so much of, and Quince Mil's residents hoped that their town was soon to be connected to this more reliable source of power. The Interoceanic Highway construction project promised not just the road but all kinds of other infrastructural connections in its wake—most notably, a reliable electricity supply and mobile telephony. For the time being, however, all they had was their faulty oil-driven system, which depended on an unreliable supply of fuel from Lima. Although the mayor insisted that her departure was to assist the restoration of power, a determination among some residents in the town to stop her leaving had drawn people to the police post where she had taken refuge from an angry crowd, who could just be made out in the blackness as pinpricks of occasional torchlight. Closer to the crowd, it became clearer that they were blocking the road, crowding out any vehicle that wished to pass with a condensing surge of bodies. People were simultaneously demanding that the mayor leave office and that she stay and talk through the problem of the electricity supply with them.

We joined the crowd. With no light and no television or radio, now that the café's generators had powered down for the evening, there was nothing much else to do, and many people, like us, were there as spectators. Other people were more passionate. The breakdown of the electricity was damaging to some businesses, and they felt that the mayor had neglected her duty to ensure this vital service. Some took the electricity breakdown as symptomatic of the fraudulent manipulation of political office—why had the spare parts, the gasoline, the necessary expertise not been paid for in good time?

As the evening lengthened, the crowd grew bigger, and by now many people were shouting. The police, using the loudspeaker on the police car, were trying to calm people down and telling them to disperse. After half an hour or so the mayor herself appeared from within the police post. She said she simply wanted to get to the city to sort things out with the higher authorities. But people shouted back that she was lying—how could they trust her? They thought it more likely that she was just trying to escape—or even abscond with what was left of the local funds. They shouted for other town council members to come and give their account. One of these men finally arrived and, taking the microphone, stirred things up further by asserting that the mayor had never supported their demands for the local electricity generator to be repaired. The engineers' report had not been acted on, and when they had asked to see it they had been told it had gone missing. At this point people started shouting for a change of mayor—indeed, for this clearly more engaged council member to replace their current illegitimate mayor then and there. After much argument, and a solid refusal to unblock the road, they got the mayor to agree to stay in town. They would have a meeting with her on the following morning where she would, in front of an audience of townspeople, contact her superiors in Cusco, so that everyone in the town would be able to hear that she was indeed making the right demands on their behalf. Only then would she be allowed to go to the city, under clear instructions of what she was to do there. When we returned to Quince Mil a week later, the electricity was back on. We were told that mayor had negotiated a delivery of oil during the telephone call she had made publicly to the relevant authorities in Cusco on the morning after the protest.

Infrastructural breakdowns like this offer interesting moments of insight into the latent politics of infrastructural form. If key aspects of the politics of infrastructural form become hidden from view by engineering techniques, whose achievement is to enable infrastructures to appear to lie uncontroversially beneath the world of human affairs, infrastructural breakdowns disrupt this work of smoothing. In doing so, such breakdowns raise powerful questions about where infrastructures have come from, who they are for, and what social effects they might be expected to have.

A year later, the construction of the Interoceanic Highway had proceeded to a point where the consortium was working near Quince Mil. The female mayor was no longer in office, and the mayor who had preceded her (and had previously been thrown out of office) was now back as the new mayor. Eduardo, who worked for the community relations team at CONIRSA, had become involved in protracted negotiations after it was reported that the new mayor was inciting the people of the town and those living in other zones to block the road in support of better compensation for the use of local materials.

CONIRSA was negotiating to establish a quarry (*cantera*) near Quince Mil. As a gesture of goodwill, it had agreed, for the duration of its use of the quarry, to pay the town a "donation" of ten thousand soles a month toward the development of the town. However, the new mayor had spread a rumor that CONIRSA was planning to rob Quince Mil, claiming that it was refusing to pay anything at all, and arguing that it should be donating at least forty thousand soles a month in return for the extraction of the raw materials. Eduardo was exasperated: "You see they squeeze us and squeeze us to get as much out of us as possible! They see CONIRSA as a big company with lots of money and they want to get as much out of us as they can."

Another open meeting was called, and Eduardo was given the chance to offer his side of the story. Once he had explained to everyone what the deal actually was, the vote came in 43–3 in favor of signing the contract with CONIRSA. The mayor yielded to the will of the people and agreed to sign. This was the end of meeting, but it was not the end of problem. Despite implying that he would sign the contract, the mayor then refused to do so, saying once again that he wanted more money. Eduardo's frustration was palpable: "If it comes to it, we can give him an ultimatum. We will let the state regulator know about his games and then they can intervene."[1] As far as Eduardo was concerned the mayor of Quince Mil had personal interests, *"intereses personales,"* that he was looking to satisfy. With elections on the horizon, these interests were both financial and political. As we moved on, Eduardo braced himself for a lengthy battle ahead.

Prompted by the regular occurrence of these kinds of confrontations, this chapter extends our discussion of the politics of infrastructural projects into a consideration of the status of road construction projects as public works. Since Habermas (1989), the analysis of contemporary political processes has acknowledged that alongside the state there lies a parallel and equally important political formation that we call "the public." In *The Structural Transformation of the Public Sphere,* Habermas demonstrated how the development of the modern nation-state was paralleled by the simultaneous development of a public sphere of rational political debate within which issues of politics were constituted as problems that should be debated by the population, whose interests the state represented and in whose name it acted. Recently, a number of anthropologists have built on Habermas to return to the often taken-for-granted figure of the public in contemporary political relations. They have done so to critically assess the role that the political idea of "the public" plays in enabling projects of social transformation, and to explore its often deleterious effects.[2]

Infrastructural projects like the road construction processes we have been studying are explicitly established as public works. By now it should be clear that the road construction projects that we followed were deeply imbued with ambitions for social transformation in the name of public good. As we illustrated in chapters 1 and 2, the dream of social transformation in Peru has been both a dream of state territoriality and a more complex struggle over different kinds of personal and social ambitions that roads both seem to offer and to withhold. As well as roads themselves having the capacity to produce a variety of political dreams, we have also explored how the micropolitical negotiations of day-to-day engagements with questions of material transformation, health, and safety and the circulation of economic resources also have important differentiating effects, participating in the reproduction of the very distinctions that these roads are expected to transcend. We come back to this point in our final chapter. We have also explored, in chapter 5, how in spite of being plans of national transformation, infrastructural engineering projects are supported by international flows of capital. They are often characterized by an ambivalence regarding how international flows of money meant for public works end up (both legally and illegally) in the pockets of private individuals located in particular institutional and political positions.

What the previous chapter also demonstrated, and the vignette with which we open this chapter illustrates, is that contemporary projects of road construction not only rest on a concept of the "public good" but also place under scrutiny the very question of what kind of public status these infrastructural projects should have. The projects are funded through a mix of public funds and private capital and involve a combination of engineering firms, state regulators, legal institutions, local political structures, and individual interests whose contractual, social, and economic relationships are often difficult to disentangle. Moreover, they enter into highly politicized spaces where populations are already engaged in negotiations with local and national politicians over previous infrastructural failures. Our aim in this chapter is not to attempt to disentangle these constantly shifting relational arrangements, but rather to explore how infrastructural projects which are characterized by these particular configurations, offer the grounds for an interrogation of the status of contemporary infrastructural projects as public works. By exploring the ways in which the *public* role of infrastructural projects gets negotiated, we will be able to further demonstrate how infrastructural projects manifest as sites of political transformation.

Infrastructure's Publics

To deal with the demands of local politicians and the requirements of local populations, both of the engineering consortia we worked with had developed explicit institutional mechanisms for dealing with what they called either "social," "cooperative," or "community relations." When the Consorcio Vial Nauta was in charge of the Iquitos-Nauta road construction project, for example, a community relations team was established that worked alongside an environmental protection officer to ensure that local populations understood what the road was for and what benefits it was likely to bring to them. The main preoccupation of the community relations team was to manage the expectations of local populations about the number of jobs the road construction process would generate for them. Although road construction historically has required a large workforce, the head engineer of the Consorcio Vial Nauta explained to us that developments in machinery meant that there simply was no need any more to employ large numbers of local people. Jobs were, however, still seen as a crucial local benefit of the road construction process, and the engineering firm recognized that to avoid a deluge of complaints it would have to manage the expectations about how many jobs would be available.

The environmental protection officer also worked closely with local populations in managing some of the effects of the road construction process. Her responsibility lay in ensuring that the environmental side effects of the road construction process would not be detrimental to local people (e.g., through contamination of water supply via runoff from the road), and in ensuring that local people did not use the connective opportunities that the road was opening up as an excuse for engaging in environmentally damaging activities. Much of her work involved talking to people living near the road and educating them about the environmental effects of deforestation.

On the Interoceanic Highway, each of the administrative sections (*tramos*) of the road had its own community relations team. As on the Iquitos-Nauta road, a large part of the work of the community relations teams was taken up with issues around employment. The construction consortium was under a legal obligation to provide jobs to local populations. As the construction process moved along the route of the road, it was envisaged that people living in different communities would be given the opportunity to work on the project. Although skilled labor was sourced from all over Peru, unskilled labor came mostly from local settlements, where people normally made a living from trade, commerce, and small-scale agriculture. Employment on the construction project not only offered comparatively good wages but also entitled people to health insurance, and it was thus seen as highly desirable for many people living in the towns and villages in the vicinity of the new road. As in the Iquitos-Nauta project, however, the community relations team on the Interoceanic Highway construction project had to deal with the fact that there were not enough manual unskilled jobs for the number of people that wanted them. Committed to a principle of fairness, the community relations teams set up a ballot system, whereby everyone who wished to be considered for employment was asked to register with the consortium, and each week names were chosen at random and the successful candidates' names displayed on a notice board.

Another key area of work for the community relations teams on the Interoceanic Highway project involved negotiations over land rights. Peruvian law states that land can be appropriated forcibly for purposes of public necessity or national security as long as the recognized beneficiary of such land expropriation is the state.[3] Road construction projects, which are developed as a "public good" and regulated by the state, are one such instance in which compulsory land expropriation can be justified. Land expropriation includes both the purchase of the land for the road itself and the use of adjacent tracts of land as either quarries or depositories for excess materials. The Peruvian state has rights to materials that lie under the subsoil, although it also has an obligation to replace the top layer of soil so that the land can be used by its owners for agricultural subsequent to the construction process.

Through these activities, the community relations teams on both road construction projects were involved in the process of establishing the road as an object of public concern. On the one hand, the purpose of the road itself was to bring general benefits to the public, and thus much of their work involved encouraging local populations to act in the public interest, for example, in giving up their rights to land in exchange for compensation. On the other hand, it was written into law that local publics, often living in impoverished communities, should be given the opportunity to benefit directly from the road construction process by being given privileged access to the (limited) employment opportunities the construction process itself generated. We might say that the road construction project, as an infrastructural project designed to institute a social transformation, necessitated the development of methods of engagement with both national and local publics.

Another aspect of the work of the community relations teams, however, was to deal with disruptions to these notions of the national and local public good. One way in which the public good of the road construction was challenged was by pressure groups such as environmentalists, who were concerned that the identification of certain national and local publics as key to the road construction process led to other kinds of social, environmental, and political positions being systematically ignored. On the Iquitos-Nauta road, for example, there were some very vocal environmental campaigners who argued that, in spite of their work on environmental protection, the construction consortium risked irreversibly damaging fragile white-sand forests, which were home to many endangered species. Similarly, on the Interoceanic Highway there was mounting concern that the road construction would increase access to environmentally destructive mining and logging activities. There were also fears that the road would enable the expansion of extractive industries into areas that were currently protected as reserves inhabited by "noncontacted" indigenous groups. In these cases issues of environmental justice were raised that connect back to our previous discussion of the rights and the agency of earth beings and environmental forces alongside the more scientific engagement with notions of biodiversity and environmental futures.

If the formal methods of justification of road construction processes depended on an appeal to national and local publics as imagined beneficiaries of the road, the challenges to these justifications might be understood as examples of what Michael Warner and Nancy Fraser have termed (subaltern) "counterpublics" (Fraser 1992; Warner 2002). Counterpublics, like publics, are reproduced through discursive engagement, and they often seem to differ from publics merely in the content of their argumentation. In a potential critique of Fraser, Warner argues that "Fraser's description of what counterpublics do, 'formulate oppositional interpretations of their identities, interests, and needs'—sounds like the classically Habermasian description of rational-critical publics with the word *oppositional* inserted" (Fraser 1992; Warner 2002, 85).

Nonetheless, Warner chooses to hold on to the notion of the counterpublic. He argues that a counterpublic is identified not just through an oppositional politics, although this is a dimension of what identifies it, but also through the nonconformist nature of its intervention: "the discourse that constitutes it is not merely a different or alternative idiom, but one that in other contexts would be regarded with hostility or with a sense of indecorousness" (Warner 2002). Warner's definition of counterpublics is useful in drawing our attention to how political debate proceeds through awkward confrontations and hostile encounters rather than through reasoned argument. However, our work with the community relations team suggested that the focus on oppositional discourse fails to address the "politics of refusal"[4] enacted by those whose modes of argument step outside the frame of established debate altogether.[5] The rumors propagated by the mayor of Quince Mil worked by the avoidance of an identifiable public voice. It is this ability to refuse without reasoned argument that we explore in more detail in the following section.

Infrastructural Counterpublics

CONIRSA's community relations team had a wide range of expertise. Some had worked in the area for many years and had a good understanding of, and close interpersonal relationships with, the more politically active members of the local communities. Others had gained qualifications in social sciences, psychology, or law. From the perspective of the general management, the community relations units were understood to be a logical place for anthropologists to spend time. Thus, we were given the opportunity to accompany the community relations officers as they engaged in negotiations with people living along the road with regard to issues of land transfer, future jobs, and the use of their buildings as sites for technical equipment.

Tom Grisaffi, who worked for a short time as a research assistant on our project, was also positioned in this way, and witnessed the following encounter. The community relations team was called to attend an incident where someone was reported to be blocking the road. When they arrived, a man called Roberto stood there, watched over by an exasperated police official. "Oh no, not him," moaned Layla, the community support officer. Apparently Roberto had been a constant source of problems. Only six months previously a laborer was working near Roberto's property on a rocky outcrop, and Roberto had pushed him off, breaking the worker's arm and leg. He had not been prosecuted because the company was doing what it could to placate the community, and charging someone with assault was not necessarily going to help to get the road built. But Roberto was trying their patience. This time he had been found lying on the track of the long cable fuses that are used to rig the dynamite, in an attempt to stop the explosions. The police had had to climb up a rocky scree and manhandle him down and then carry him far out of harm's way. Emilio, the head of the community relations team, got out of the car and talked to him in a friendly enough manner in Quechua. It turned out that what Roberto really wanted was a job for a member of his family. His four children were all grown up and already employed, but his two nieces needed work. Emilio eventually agreed to contract one of the girls, but only on the condition that Roberto stop "*jodiendo*" (fucking things up). The girl's employment was dependent on Roberto's good behavior. A deal was struck. If Roberto stepped out of line, Anita, his niece, would lose her job. Layla gave Roberto a slip of paper confirming the arrangement. Later on in the car Layla told Tom that Roberto is *loco* (mad). Although Tom could see what she was saying—Roberto did have a bit of a gleam in his eye—at the same time he reflected that his actions had been effective. He had, after all, achieved what he wanted—a job for his niece.

From the perspective of the community relations team, the incessant stream of demands for different kinds of compensation or payment was a source of great frustration. It would have been convenient to dismiss these demands as individual and not political actions, but the sheer number of demands, and the forms of community organization that they often generated, made this distinction difficult to maintain. In an attempt to respond to the desires of local communities, the community relations officers tried to use all kinds of public meetings to explain to people the limits of the consortium's responsibility, and to indicate the formal mechanisms of public consultation and recruitment through which people's opinions and requirements would be dealt with in a fair and measured manner. Nonetheless, public meetings often ended up further polarizing the engineers and local residents. When, for example, the residents of a small Andean town argued that they wanted the highway to pass straight through their central square, and not to go via a bypass, their suggestion was dismissed by the engineers as patently ludicrous. Who in their right mind, they asked, would want a highway plunging its way straight through a village square? In the mechanisms that existed for public engagement where discussion was meant to take place, positions that were deemed unreasonable were often overruled and ignored. Some were dismissed as irrational, while other positions, like Roberto's, were simply deemed "mad."[6]

At the same time, these actions on the part of local people and communities were legible as demands that the community relations team could either engage or ignore. The conversations were awkward and disjunctive, but the conversational format produced a recognizable demand, even from a disruptive figure such as Roberto. However, we observed a number of incidents during our time with the community relations teams that posed a much more difficult challenge to those who were responsible for ensuring that the construction project operated as an ethically robust, fair, and equitable public work. It is to some of these even more awkward encounters that we now turn.

Impossible Publics

During the time that we spent with the company that was constructing the Interoceanic Highway we were introduced to Delia, the lawyer from the Ministry of Transport and Communications. Delia's job was to ensure that the people who were required to sell their land for the project were properly represented and informed of their rights, and that they filled out all of the paperwork needed to process their compensation claims. While we sat in the back of a van on our way to one of the villages that lay along the proposed route of the road, Delia told us, exasperatedly, about a problem in recent negotiations in one of the communities. One particular man was refusing to sign the paperwork to hand over his land to the Ministry of Transport and Communications, so Delia had gone down to investigate. She spoke to the man, Aurelio, and asked him why he wouldn't sign. He replied, "*no me da la gana*" (I am not moved to).[7] Despite the fact that the road was to cut directly through his land, Delia explained that she could not elicit a reason that she could work with for his lack of cooperation. She told him that he needed to give her a more concrete excuse and suggested that he could argue that this land was his only livelihood, and from there they could begin to negotiate with the Ministry of Transport and Communications about compensation. Aurelio simply replied "no," and once again said "*es que no me da la gana*" (I am not moved to). Delia told us that she was frustrated by his refusal to cooperate, particularly as she felt he was actually getting a rather good deal. The road was going to pass only through a portion of his land, but she was willing to argue that he should be paid as if all of his land would be used.

Delia told us that in the end her only option had been to turn to Aurelio and say to him, "Fine, we will get legal documentation together with signatures from the whole community, and we will write on that document that Sr. Aurelio Camargo will not participate because 'he is not moved to'!" Aurelio apparently laughed at this suggestion. "You can't put that!" he remonstrated. "So what should I put?" Delia asked. "What was that other thing you suggested?" he had asked her. As she laughed at his audacity, he said, "Oh, go on, then, I'll sign 'because you've got a pretty face.'"

This interaction seems to open up an important disjuncture, or gap, in a narrative of publics and counterpublics that we have so far suggested characterizes the politics of public works. Delia's frustration with Aurelio was not due to his mobilization of a counterpublic position in the ways in which we have seen in the examples of opposition to the actions of the road construction consortium over jobs or routing. In fact, it was Aurelio's failure to articulate either a public or counterpublic position that appeared to render his response so infuriating to Delia in her role as a public servant. Aurelio's statement that he was not moved to sign the papers that he was presented with thus seems to capture a different kind of engagement, one that fails to cohere within the contemporary politics of publicization (Jimenez 2006) that we have described so far.

Intrigued by the gap opened up by Aurelio's refusal to sign, we began to wonder about how infrastructural projects generate the conditions that draw out this form of public (non)participation. We soon became attuned to other instances where there appeared to be a similar disjuncture between a politics of public or counterpublic engagement and a mode of acting that was based on a practice of apparent disinterest or refusal to engage with the basic terms of these public works projects.

When two scientists from Arequipa visited the engineering camp at Ccatqa on the Interoceanic Highway to make measurements of air quality and noise levels, they faced an ambivalence toward participation on the part of local residents that they said they were well used to. Their measurements were for the benefit of the local community, a legally stipulated check and balance on the company to ensure that air and noise pollution did not reach excessive levels during the construction process. To properly measure the effect on the community, the scientists had to find secure places in the village to place their measurement equipment. To generate the correct measurements, they needed to find a house that was the right distance from the camp and was occupied overnight so that the expensive mobile laboratory they planned to install would be safe. Walking around the village with the head of the community, they knocked on the doors of several houses before finally finding one where the owner was in. The owner opened the door cautiously and peered out suspiciously at our group—two scientists, two company reps, the head of the community, and Hannah. The scientists asked if he would mind if they used his house for their measurements, and the man looked very unsure. Assuming that his uncertainty was for monetary reasons, the engineers said that they would compensate him for any electricity used—giving him ten soles (about £2) even though the cost of the electricity would only be about three soles. This token monetary gesture did not appeal to the man at the door, though, and he shook his head, saying very little but retaining the suspicious look on his face. Realizing that they were not getting anywhere, the group moved on to find a more cooperative participant who was willing to help the cause of knowledge collection for the public good.

The actions of Aurelio and the man who opened the door to the group of scientists in Ccatqa suggest a response to the power relations implied in transformations of the material world that is very different from the energized mobilizations against the rationality of road building produced by environmental campaigners and organized community groups and their appeals to public sentiment. Rather, in these moments of hesitation or aversion, these actions seem to allow the appearance of a third space, an alternative position that is outside the terms of discussion on which the politics of public accountability are posed. We can neither argue that they are reproducing a model of public accountability on which infrastructural interventions are premised nor that they are resisting these interventions through an oppositionally driven remobilization of the social conceived of as counterpublic. For this reason we choose to call the positions they produce those of "impossible publics."

The Agrammaticality of Impossible Publics

To try to find a way to think about these impossible publics we turn briefly to Herman Melville's 1853 story *Bartleby the Scrivener: A Story of Wall Street.* Melville's short story has been the subject of considerable philosophical interest in recent years,[8] and it offers us a philosophical perspective on the politics of refusal. The story centers on Bartleby and his dogged refusal to engage the wishes of those around him on the grounds that he prefers not to. This refusal is not definitive; he never says that he won't, nor that he will; he simply prefers not to. Deleuze (1998) suggested that Bartleby's refusal to engage in reasonable explanation might be read as a form of *agrammaticality*, a logical statement that is nevertheless incomprehensible or at least awkward in relation to normal grammar. Understanding what Bartleby says leaves his interlocutors with no understanding of how to engage with him. This is because Bartleby's statement is, as Bartleby himself says, "not-particular" (it is not a statement of will or of refusal in relation to specific situations); it is a response driven by its own internal logic, rather than a result of the conditions of the external acts to which the statement refers. It therefore posits a systematic social disengagement, while simultaneously maintaining a strict adherence to the very social norms of logic and politeness that render his activities nonconfrontational.

The story of Bartleby helped us to think about how Aurelio's response to Delia's request placed him outside the habitual binarisms of domination/ resistance to which political argumentation most commonly refers. Deleuze suggests that Bartleby's repeated insistence that he preferred not to act in particular ways was a form of social disengagement, but this seems less pertinent to our discussion. Far from being nonparticular, we suggest it is much more likely that the responses of Aurelio and the people who spoke to the scientists in their houses were responses to particular interventions into their lives. They were being asked to become incorporated in a project of infrastructural transformation characterized by complex and often opaque relations between corporations, the state, and international financial capital. By the time these moments of engagement take place, there is little room for maneuver for people who are being asked to participate personally in the name of a generalized public good. As we noted earlier, when it comes to land expropriation in particular, the construction consortium ultimately has the full force of law behind it to ensure that the road is built in the name of the public good. Agamben's (1999) suggestion that Bartleby's statement is not a statement of will (either his or that imposed by another), but a statement of potentiality, is more helpful. Crucial here is the notion that Bartleby's statement is *not* a refusal per se. The possibility of future engagement is left open as a potential. Similarly, at some point Aurelio might be moved to sign, but of course it is equally likely that he will not be.

Agamben suggests that "our ethical tradition has often sought to avoid the problem of potentiality by reducing it to the terms of will and necessity" (1999, 254). This suggestion that potentiality is specifically *not* about will is helpful for thinking about Aurelio's statement—that he is not moved to sign—in a way that takes us beyond the opposition of positions of domination versus resistance. If Aurelio's statement was a willful claim, he would easily be able to be incorporated into the ethical schema by which his opposition to the road was understandable *as* a statement of opposition. The next step would be to rationally discuss appropriate levels of compensation, or articulate the grounds for rerouting the road, or both. This option was successfully displayed by Roberto in his search for a job for his niece. What Delia required was precisely a statement of will from Aurelio in order to activate the compensation process or to make a clear case for rerouting. The fact that he was not moved to sign, however, kept open the question of whether or not he opposed the road going through his land. The potentiality of his position rendered Aurelio logically unincorporable into the ethical tradition that allows state development projects to proceed in conversation with its publics on the assumption of a model of equivalence that allows the meaningful (i.e., reasonable) compensation of one good for another—in this case, of land for money, or, in Roberto's case, of a job for his niece in exchange for his good behavior. It is in this sense that we suggest Aurelio, and the man who was reluctant to agree to letting the scientists install their equipment on his rooftop, move us to consider a position that we have chosen to call that of "impossible publics."

Anthropologists have tended to find their departure from established theory not through literature or the thought experiments of Western philosophy, but through engagement with the perplexing character of social practices that appear in the course of ethnographic fieldwork. Here, the ethnographic mode of attention requires a reversal, whereby the strange is made familiar and the familiar made strange, in order to explain the moment of confusion that comes from the disjuncture between different modes of acting. Much has been made of the ways in which anthropologists draw their analytical apparatus out of the interplay between the containments of theory and the extensivity of ethnography.[9] Ethnographic theorizing tends to proceed not through the imposition of Theory with a big "T," but through practices of redescription that retell what at first sight appears unfathomable and, in the process, render it comprehensible.

In this respect, to simply cast the position of what we have called "impossible publics" as a philosophical truism tells but one side of the story. The other side appears when we turn our attention to the way in which the actions of these impossible publics can be explained ethnographically in terms of the specific contours of the events in question. We suggest that one way of opening up a theoretical discussion on the politics of public participation in engineering projects is by reconsidering these moments of encounter not as a moment of aporia in a general process of "public" engagement—a position that renders them "impossible"—but rather as forms of "engagement" that posit alternative modes of publicity.[10] This move alters the terms of the question, from one concerned with people's capacity to participate in a predefined version of the public good to one concerned with the terms within which people's actions might be seen as a form of participation or engagement at all. In the following section we explore this further, by introducing the idea that impossible publics are not so much incommensurable with an infrastructural logic as they are rather "otherwise engaged" (Harvey and Knox 2008).

Otherwise Engaged

To explore what we mean by "otherwise engaged," we return to Roberto and the criticism that he was in some senses "mad." Even though his methods of gaining the attention of the engineering company were perhaps unconventional, his expectation that the road-building process should be able to generate for him and his family some tangible benefit was not unusual. The scientists who were installing the equipment on the roofs in Ccatqa recounted the story of a man they met when they were working on a mining project. He lived near the mine and had agreed to the long-term installation of monitoring equipment on his house. During their first visit, the scientists had negotiated compensation of thirty soles per month, but when they returned to check the monitors a month later, the house owner said he had changed his mind and now wanted sixty soles of compensation a month. With a month's worth of data collected, the scientists could not take down their equipment and move it elsewhere, and so they conceded to this request for more money— only to find that, the next time they went back, the man was demanding that he be paid 120 soles a month. Negotiation over the fairness of this claim rested on balancing the interests of this man as an individual affected by the investigations of the scientists with the appropriateness of different levels of compensation for studies in the public interest. What the scientists were struck by was the incapacity of the man in question to recognize the appropriate scale within which his compensation claim would be read as reasonable.

In recent years anthropologists have become increasingly interested in the ways in which the people who have become subject to the environmental effects of large extractive or infrastructural projects negotiate the tricky business of making compensation claims on large multinational corporations.[11] These claims often appear way out of scale with the expectations of the transnational corporations as to what would constitute a reasonable request for compensation. Kirsch (2006), for example, writing about the attempts of Yonggom speakers in western Papua New Guinea to gain compensation from the Ok Tedi Mining Corporation, has discussed how compensation claims have a tendency to increase in scale and scope in ways that are frustrating for mining companies and the state. He introduces the concept of an indigenous "analytics" to show, from the Yonggom point of view, how they make sense of and engage with the environmental destruction produced by a large mine located upstream of their villages. Kirsch associates the endless extendibility of their claims on the state with a Melanesian form of sociality that privileges practices of association rather than practices of detachment.[12]

The problem with approaches such as that outlined by Kirsch is that even though they provide a fascinating account of situated practices of engagement with large-scale infrastructural or industrial projects that impinge on people's lives, they have less to say about the way in which the analytical practices of indigenous groups might themselves have been constituted in relation to the very processes they seek to analyze. In the case of compensation claims on the roads we studied, we suggest that to understand their apparent incommensurability with the logic of infrastructural transformation, we need to consider the way in which people making these claims were negotiating a particular form of engagement with the anomalies inherent in the appearance of infrastructures in their lives.

As we outlined in the last chapter, people at all levels of the road-building process find themselves caught in a tension between acting simultaneously in the public and private interest. We explored how corruption narratives capture the way in which, in spite of the self-proclaimed transparency of public works projects, their organization produces the conditions within which ambiguity, secrecy, and inequality proliferate as a way of enabling private gain to accrue in the guise of public concern. The people who we have termed "impossible publics" are themselves highly attuned to the difficulty of acting in the face of the powerful networks of relations that hold infrastructures in place. As we described in chapter 2, road construction itself is in many respects a conjuring trick (Tsing 1999) whereby land is transformed into the fetishized dream of a future state-of-being. If politicians, engineers, and officials appear to gain personally from the road-construction process by creating spaces of practice that escape the opposition between public and private interest, it is perhaps not surprising that the people along the road do the same. In this respect we might see the unfathomability of the actions of impossible publics as a kind of mimetic mode of relating to the obfuscations, slipperiness, and ambiguity of the relations that go into making infrastructures a bedrock for society.

The very term "infrastructure" points to the effect of fixing things in place. In spite of the way in which these roads are established as projects of social transformation, we have shown that they are also manifestations of social, legal, and technical methods that work to fix relations and allow them to be held in place over time. This achieved fixity, we suggest, also has powerful effects in generating the conditions of possibility for what kind of political action is possible. As we have shown in earlier chapters, technical processes work to displace social relations. The community relations teams of the construction companies pick up the displaced question of the social effects of material transformations and work hard to develop techniques through which both national and local publics and counterpublics can be given a voice in the engineering process. Nonetheless, people are often well aware that the possibilities that this space of public consultation holds for social engagement is limited. By the time they are consulted, so many objects, institutions, and legal relations have already been fixed that the incorporation of public opinion into this arrangement is limited in its effects. Occasionally, however, people find ways of intervening directly in the networks of relations that are required for the infrastructural projects to endure. The man who kept upping his compensation claim could do so because the scientists had already invested time, energy, and resources to establish data that was essential for the project to proceed.

Lest we appear to be giving too much to the capacity of individuals to stop the power of a large state and corporate engineering project, we end with a final observation. When Penny returned to the town of Ocongate after eighteen months back in the United Kingdom, the place that she had known for over twenty-five years had undergone an incredible transformation. What had been a fairly isolated town some eight to ten hours away from the city was now only a two-hour journey from Cusco. Everyone was buzzing with talk of their most recent confrontation with the company. Stories were circulating in Cusco that the people of Ocongate had staged a mass demonstration at the construction camp that led to violent clashes, and there were even rumors of local people being "disappeared."[13] The new alliances that had been struck up between the company and the state were confusing. The engineering consortium had signed an agreement with the police to protect the works, a somewhat ambiguous situation that once again draws attention to the indeterminacy of the public interest. State employees were required to protect the corporation in the public interest as opposed to the supposedly private interests of the counterpublic collectives who were protesting. The camps that we had lived in eighteen months before, which we had experienced as relatively relaxed spaces, were now controlled and patrolled by armed guards. Back in Ocongate it emerged that the rumors of conflict were related to a demand for compensation that local people had made of the consortium.

Ocongate was one of the towns where it had been decided not to drive the highway through the main square but to route it along one side. As a result of this decision, which went against the wishes of many people in the town, people were left feeling confused and dissatisfied. In spite of all the promises that the road would transform their lives, in fact the basic infrastructures of their lived spaces hadn't changed in any noticeable way. As a way of recovering some tangible benefit for the town, they asked, by way of a good-will gesture, that the stretch of road in front of the school might be asphalted. There was too much dust there, and they worried that it was bad for the children's health. Why, they asked, couldn't the company just use its machinery, its labor power, its materials—it wouldn't take much, and it would be a mark of recognition and an engagement with the lack of fit, or the limits to equivalence, between the different kinds of public good that roads were supposed to produce. The company, however, were adamant that this was not its responsibility. Company officials argued that they could not engage in public works of a general kind, pointing out that this is what local government is for, particularly under conditions of decentralization when local government is given money precisely to spend on those projects they deem of greatest importance. The consortium argued that local people should organize themselves, decide what they want, and use locally available monies to get it done.

The director of the school, who was frustrated by the impossibility of generating a meaningful conversation with the company over this issue, then sent a letter to all the parents requiring them to stage a protest. He said that if they didn't come they would be charged thirty soles (for many this could amount to two days' wages), and if they arrived late they would be charged ten soles. The parents, numbering some 750 people, dutifully turned up and marched on the camp, sweeping up the mayor of the town on the way. Company officials sent a representative to tell them that they were waiting for the head engineers to arrive and meet them. In fact, taking advantage of their new proximity to Cusco, and drawing on the new agreement with the police, they had called Cusco for protection. Two hours later, eighty riot police arrived to disperse the crowd with batons and teargas. Men, women, and children, who had largely turned up because they couldn't afford the fine for nonparticipation, were battered and intimidated. Luckily, nobody was killed. The police spent the rest of the day patrolling the village, arresting people and imposing order.

The school director and the mayor were both prosecuted by the state for inciting the confrontation. It certainly appeared that the school director did compel those people to whom he had access (the parents) to attend the demonstration. However, he was also responding, as he saw it, to an issue of public concern by orchestrating a public protest. The mayor had tried negotiating with the company by trying to act as the appropriate representative of public concern, but the demands the public was making were considered unreasonable. Voicing unreasonable demands, but with no real intention of constituting themselves as a political opposition, the parents were then utterly horrified by the company's response because they did not see themselves as doing anything more than asking for some acknowledgement of the gaps in a public consultation process they felt had deceived them, producing a public work that failed to accrue any benefit for themselves. Thus, the demand, from their perspective, was not an inappropriate request to the company for public works, but rather it was some reparation for their disappointment, their need to see something change in their village in direct relation to the road-building process.

Given the contradictions and ambiguities that people like the residents of Ocongate face as they work to find a way of participating and engaging in road construction processes, it is understandable why the other means of engaging this process that we have recounted in this chapter exist. Rational debate is often a dead end, and counterpolitical protest is a risky strategy. Acts of invisibility, intransigence, or obnoxiousness, on the other hand, are less easily contained than more explicit forms of public protest. Moreover, these are precisely the methods often used by those who appear to gain some personal benefit from the road. As we saw in chapter 5, in embarking on projects of infrastructural transformation, engineering firms, capital investors, and politicians constantly seem to slip between visibility and invisibility, obstinacy and compliance, and transparency and secrecy.

Conclusion

We return, in conclusion, to Aurelio and his refusal to sign, which prompted the writing of this chapter. Although we cannot know for certain his reasons for not wishing to sign the land transfer document, his statement has nonetheless provoked us to consider more broadly the place of impossible publics in processes of infrastructural development. Aurelio's initial statement directed our attention to a complexity unacknowledged in the opposition between publics and counterpublics as the discursive modes through which a political philosophy of politicization comes to be articulated. The subsequent ethnographic examples enabled us to acknowledge some of the intersecting lines of tension that traverse the politics of public consultation. Whereas community relations officers have to find willing participatory publics who can be consulted and called on to debate the terms of their cooperation, those who find themselves on the side of the road occupying the position of these "found publics" are required to tread a delicate line between a need to act rationally, the possibility of keeping open their own potentiality, and the threat of falling into positions where they are typecast as uncooperative and obstructive individuals. That their responses have the effect of producing an ambiguity as to what interests they are serving is perhaps unsurprising when we consider their actions as a form of mimetic relationship to an economic and political system whose method of functioning is often experienced as equally obscure.

We have suggested that the refusal of a public voice in these cases is not a refusal of a political position. Rather, we argue that in these encounters we find the political reconfigured as a mode of negotiation where what is important is not only the rational discourse of Habermas's (1989) publics but also a yet-to-be-determined capacity to navigate the vagaries, complexities, and intransigence of infrastructural relations. By focusing our attention on the appearance of what we have called "impossible publics" in our ethnographic work, we have opened up the notion of the public to acknowledge the role of what Deleuze calls a minority politics. Impossible publics articulate, through practices and positions that appear incommensurable as dimensions of the public sphere, the very contradictions that we have shown in this chapter to be inherent facets of projects of infrastructural transformation. The impossibility of these publics derives not from a culturally different mode of explanation or analysis, which is somehow misunderstood by Western rational discourse, but from the very contradictions that the separation between a realm of material transformation and a realm of public engagement itself puts into play.

Chapter 7

Conclusions

Inauguration, Engineering, and the Politics of Infrastructural Form

Our study of Peruvian roads began in response to the challenge of how to work ethnographically on the state. Roads offered us a particular way into thinking about state presence. As material infrastructures that are planned, executed, and owned by the state, roads also demonstrably bring that state into being, creating and recreating its territorial form and enacting its paradigms of ownership and control. At the same time, as we have shown, roads also exceed the state as they become part of the mundane material fabric of people's lives, producing possibilities and limitations that go beyond any specific plan for integration, connectivity, or even abandonment. In previous chapters we have traced the material politics of road construction with respect to two highways. The histories and contemporary ethnographies of these roads have offered a perspective on some of the ways in which the modern Peruvian state emerged, in practice, through uneven flows of people, knowledges, materials, money, and other less tangible cultural resources such as hopes and fears, beliefs, and memories. We also found the "state" to have been a somewhat discontinuous presence in most people's lives, confirming our sense of the problematic nature of the state as a starting point for an ethnographic account of political life in Peru. Our focus on the roads, however, has allowed us to rethink the political from the more grounded, experiential, and immediate space of infrastructural formation. As we tracked the ways in which these roads were made, and attended to what they were made of, we were also able to follow the social forms that roads brought into being—the territories and displacements, the diverse forms of expertise and analytical attention, the movements and blockages, the social orderings and the transgressions.

While approaching roads as important sites of politics, we have explicitly attempted to go beyond those accounts of road politics that privilege the role that roads play in political struggles between state forces and protestors. The capacity of roads to channel and focus political confrontation is important, and both of our roads have been used for this purpose on many occasions. However, this book has been more concerned to extend the reach of the political beyond the established understanding of liberal politics—whereby the political refers to the negotiations or struggles between subjects, or between state and citizen. Our focus on roads has been a deliberate attempt to extend the space of the political to include both nonhuman subjects (forces, materials) and the refusals of impossible publics whose actions tend to be illegible to those who limit the political to particular modes of struggle and contestation.[1] The political in our usage has referred to the relationships in and through which heterogeneous forms of social difference are enacted. At times such differences emerge through conflict, argument, negotiation, or settlement. At other times they remain as implicit, unsettling forces. The tracing of how difference is articulated has thus lain at the heart of our project.

By way of a conclusion, we move our attention from the making of a road to the means by which road projects are brought to a close. Following the making of a road taught us a great deal about the ways in which road construction projects effect a politics of differentiation, particularly through the interplay between devices of standardization and unruly or unpredictable environments. We have paid attention to the ways in which road construction projects deploy and consolidate an image of a singular, abstract national territory or state space[2] and appeal to the standards and universals that underpin the authority of expert knowledge.[3] At the same time we found a constant engagement with the awareness of such spaces and knowledges as inherently open, mutating, and responsive to the dynamic environment of which they form a part. Nonetheless, from the perspective of the construction companies roads *are* eventually completed. In the registering of their completion there is a shift of perspective from the road as a living, growing relational entity to the road as an altogether more singular thing, a material form that engineers might perhaps take credit for. Indeed, many engineers talked of their hope and expectation that their works will have some kind of monumental future, standing as testimony to their achievements in stabilizing recalcitrant environments.

Inaugurations

An important moment in the enactment of this closure is the inauguration ceremony, a ritual in which politicians forge visible associations with new infrastructural systems, delivering the valuable electoral currency of public works to the media and, through them, to the electorate. Unfortunately, we never directly witnessed any of the many inaugural events that took place at various times on our two roads. They always seemed to happen suddenly, when nobody was really expecting them. We were certainly not the only ones to wonder why we never knew when these ceremonies were going to take place. Indeed, one of the most striking things about such ceremonies was their seeming capacity to exacerbate a sense of differentiation between a local population and a distant and external centralized state.

Don Emilio, one of the Nauta pioneers referred to in chapter 1, told us of how he missed the inauguration of the Iquitos-Nauta road. Emilio was among those men who had first worked to open a path through the dense forest that lay between Nauta and Iquitos. He, along with fellow workers from that time, had been informed of President Toledo's visit by officials of the regional government. This was exciting news indeed. Over the years these men had seen the road grow from an extremely local project to the public work (*obra*) that had been realized through extensive networks of collaboration that had brought international finance capital, engineering expertise, and political controversy to the small town of Nauta. The pioneers were proud of the crucial role they had played in the initiation of this project, and they, more than anyone, were aware of how difficult it had been to elicit the support of the central state. It was a local politician, the mayor of Nauta, who had originally contracted the pioneers to find the route that would connect them to the road that already existed between Iquitos and the Army camp at kilometer twenty-one. In their understanding of the process, the road had emerged thanks to local commitment and energy. It was local people, the local politicians and the pioneers, who had brought this road into being. They had opened the route. They had risked their lives, the dangers from snakebites, the deprivation that they endured for twenty-two days with only minimal food and water. It was their scout who had found the route, their *trochero* who had opened the original path with his machete, making it possible for the rest of the team to follow. The original expedition had included a technical expert who had brought a theodolite to survey the land. But, in Emilio's account, this technician was not, in any stretch of the imagination, in charge of the project. On the contrary, he succumbed to exhaustion and was utterly dependent on the pioneers for simply getting back alive. The pioneers were never paid for the work they did, and there was some resentment that despite their fundamental contribution they had received neither pension nor recognition. In fact, Emilio was at pains to point out that "things get done around here because there is no support from the authorities. They don't help you, so you do things for yourself."

Despite these decades of struggle and disappointment the men were excited by the possibility that there was finally some hope of public recognition in the form of an inauguration ceremony. They saw this event as an opportunity for the state to finally recognize the important part they had played in producing the road. For them, the road was testimony to their strength and resilience. They had never given up, but had managed to coax this stretch of asphalt into being despite the long history of abandonment and the intermittent attempts by state authorities to plunder resources and engage in flagrant acts of theft and corruption. The current head of state, President Toledo, had not been implicated in this murky history, and thus he offered a new opportunity for the pioneers to forge a productive relationship with the central state. The state might deliver on the promises of care and social provision that all politicians make in their election campaigns, where public works feature strongly as the best way to encapsulate the desire to meet local needs while responding to wider national and international aspirations for economic growth.

Unfortunately, things did not turn out well. Toledo was due to arrive at kilometer five, and the pioneers had been instructed to wait in Nauta for somebody to come and collect them. They had assumed that they were to be treated as honored guests, but when nobody appeared they finally decided to take the initiative and hired a mototaxi to take them to the ceremony. When they got there, they were not allowed through the security cordons. By the time that word had got through to those who could have afforded them some visibility in the proceedings, Toledo was no longer there. He had arrived and left in a helicopter, as presidents do. The gap between these men and their president gaped before them. Even when in the vicinity, drawn there by the precious road in which so much hope had been invested, the connection between local aspiration and central state preoccupations had once again failed to materialize.

Our absence from the event prevents us from even beginning to imagine an ethnographic description of the kind that Max Gluckman produced in his famous analysis of the inauguration of the bridge in late 1930s Zululand (Gluckman 1940). Gluckman's engagement of the "social situation" established the importance of such events for the analysis of rituals as moments of potentiality, as encounters when social and political differences are produced and reproduced as much in the erasures of others as in their affirmation.[4] The ways in which Emilio was forgotten, his inability to cross the security cordon, and his ongoing sense of both hope and disillusionment were, on this occasion, familiar but unexpected emergent effects. Just as in Gluckman's example, structural inequalities were enacted and embedded in the routine dynamics of the ceremony. Standing at kilometer five on the Iquitos-Nauta road, President Toledo could celebrate the new connectivity that the road stood for, while also reproducing the finely differentiated social space that distinguished those who were there from those who were not, and that shaped how they were there, how they arrived, and how they left. The disappointment at the heart of Emilio's story brings home how the inauguration of these infrastructural spaces routinely launches new futures that are from the outset highly problematic in local terms. In the first place, they are enacted and imagined at a scale at which local interest is irrelevant, superseded, or transcended by the force of the *obra,* which may be located but is no longer local. If the president has come to open a road, he has come in the name of a national project, and most usually to celebrate new or enhanced international connectivity.

On the one hand, then, the inauguration ceremony was a moment in which local residents sensed the gulf that existed between themselves and a centralized state. At the same time the reverse was true. Not only did the ceremonies enact the centralized state as distant and uncaring from the perspective of local people, they also produced the local population as a challenge for the state itself. The ceremonials that surround an inaugural event do not simply mark the road as a state space, but also suggest that it is a space gifted to local people. Local people, now in receipt of this gift, are obligated to make something of it: to use it appropriately, to conjure prosperity from it. In the act of inauguration local people are thus handed the responsibility of improving their lives. This obligation or sense of implicit debt to the state is interesting to observe as an effect of the politics of decentralization, and of neoliberal economic policies more generally, and reminds us of the continuing centrality of transcendent power to regimes that otherwise revolve around a sense of distributed force. Somewhat surprisingly, this sense of obligation is reminiscent of the huge public ceremonies that the Inka rulers would orchestrate to inaugurate the growing season. In these ceremonies the Inka asserted that all productive capacity emanated from their gifting to their subjects, of land, water, and ultimately life itself. The ceremonies were huge theatrical enactments of subjection orchestrated by an imperial power with an overwhelming capacity for material and symbolic control of its territories. Of course, the compulsion to participate is orchestrated differently in contemporary times, and yet the continued power of the state to define the public good, and to act in the name of this abstraction, continues to confound possibilities for local people. Furthermore, the state's capacity to invoke the public good at a scale with which specific people struggle to identify further condemns such people to marginal spaces of poverty and anger, in which they are increasingly either ignored or criminalized should they orchestrate dissent.

Following Gluckman, then, we can see that the inauguration of contemporary infrastructure projects performs a similar politics of structural differentiation to the one he so memorably described. Inauguration ceremonies enact a differentiation of the local from the national, of the state from the citizen, and of the politician from the public. However, although Gluckman's account provides a rich description of an inauguration ceremony not dissimilar to those taking place along our two roads, it misses a further form of differentiation that we have been at pains to explore in this book: the differentiation between Politics with a big "P" and the politics of technical relations.

Inaugural ceremonies are not only moments for indexing the relationship between politicians and publics; they are also taken as important opportunities to reclaim the roads from the space of technical expertise in which they have become embedded in the course of their construction. Public works are always more likely to produce enduring associations with the politicians who secured the funding than they are to elicit tributes from, or even the memories of, the engineers or the workforce who constructed them. Both of the roads that we have described were inaugurated by presidents who sensed the electoral (or legacy) advantage of a close association with these promissory spaces. Although we tend to think of inaugurations as occurring at the end of a construction project, we were struck by how often an inauguration would be announced during the middle of a project, to coincide with political events. Just before the 2006 general election, for example, President Toledo opened the first kilometer of the Interoceanic Highway—a tiny stretch of road that was completed and inaugurated before the rest of the road even had a full engineering profile. He and his ministers had worked to get the Interoceanic project off the ground, and they wanted recognition for this achievement in the electoral campaign. Toledo's successor, President Alan Garcia, was on hand to take credit for the subsequent construction. In the final weeks of his presidency, Garcia inaugurated the Billinghurst Bridge in Puerto Maldonado before it was actually completed.

These inauguration ceremonies thus appear to have little to do with the completion of the technical process of constructing a road. Indeed, the inaugural moment effectively puts *process* aside to celebrate the product.[5] Recalcitrant materials, disputed hierarchies, and unstable alliances are not discussed, nor are the negotiations over relative expertise, spheres of jurisdiction, mutual responsibility, and the ethics of environmental engagement. More unlikely still is there any reference to the multiple, and incompatible, aspirations for future transformation. After all, inauguration ceremonies are not the place where you would expect to find discussion or debate around the nuanced configuration of the public that is enacted in public works.

The power of inauguration ceremonies to enact big "P" Politics, while erasing the politics of technical relations, is suggestive of a more pervasive doubling within the politics of contemporary infrastructure projects. Take, for instance, a very dramatic instance of infrastructural politics that was brewing at the time of the final inauguration of the Interoceanic Highway. The central government, anxious to continue its collaborations with Brazilian capital and to further its initial steps to find new inroads into Brazilian markets, had begun to draw up plans for the construction of several large reservoirs. The idea was to sell both water and electricity to Brazil. What was shocking was that the reservoirs were to be built at Inambari, across the very area through which the new Interoceanic Highway now passes, destroying the road in the process. Beyond the possibility that the road might create a more integrated national territory, or facilitate flows of goods and persons—extending the networked integration of Peru with Brazil, and even creating the alluring promise of a gateway to eastern markets—the road construction project had opened up new possibilities and new relations, and a new future in which the road itself was really not that important. From the perspective of the political performances enacted at inauguration ceremonies, this calculation seemed absurd—why destroy a thing that held such political promise? However, from the perspective of our analysis—one that has attempted to recover a more complex terrain of social and technical relations at play in infrastructure projects—the plan for the reservoir is less surprising. Simultaneous to the performative inaugurations of this construction project, the government had begun to focus on how the reservoir project might be more lucrative and secure better relations with other countries. Like any infrastructure project in its early incarnation, the reservoir was shrouded in uncertainty. Indeed, the project had been suspended when we left the field, although not canceled, but this did not mean it was without its social effects. With the announcement of the reservoir, the plans and calculations of state officials had caused Inambari to become a site of considerable social conflict that involved alliances and tensions between local residents, engineers, and concerned environmentalists. People of course adapt—as they always have—but such adaptation frames the struggles of everyday life as the new opportunity threatens in small or overwhelming ways to destabilize previous adaptations and precarious livelihoods. Our work suggests that those who wish and need to understand the dynamics of such conflict and processes of adaptation would be well advised to approach the field with an awareness of the wider infrastructural politics already in play.

In these contexts of uncertainty, when roads become reservoirs, and when infrastructures turn out to be capital assets,[6] and not routes for terrestrial integration through the flow of goods and people, inaugurations index a very particular kind of political "promise." This is not a promise of infrastructural provision, for the inauguration requires that the work of construction is assumed to have already been done by the multiple actors who have effectively brought the material infrastructure to a point of (provisional) completion. Inaugurations instead gesture to a potential future that still requires work but holds promise and allure in relation to particular (but underspecified) future imaginaries: greater wealth, greater health, greater connectivity.[7] Inaugural ceremonies render technical work invisible, at the same time as they register the uncertainty of the social and acknowledge that more is needed for the infrastructure to deliver that which is expected of it. Inaugural ceremonies mark new beginnings under good omens and seek to render the works auspicious. But auguries also mark uncertainty, and there is a tacit acknowledgement that while the project might be successful, it might also fail. The handing over is the passing on of this responsibility, and a gesture that also points to the dark side of the promise of a better future, with the reminder that this future comes without guarantee (not even, as we have seen, the guarantee that the state will continue to back the infrastructure after its inauguration). As Keck and Lakoff (2013) have pointed out, the augury is above all a "figure of warning." When an infrastructure (or a single kilometer of tarmac surface) is handed over, the quality of an uncertain future hangs in the air.

The Emergent Politics of Engineering Practice

Having demonstrated that inauguration ceremonies are not primarily rituals that mark the completion of an infrastructure, but rather political punctuations in a longer trajectory of infrastructural emergence, we now turn our attention to the way in which the technical work of engineering a road enacts the closure of a project and the requisite transfer of responsibility from engineer to politicians and ultimately to the public. As we discussed in section 1, both our roads had their origins in the 1930s in Peru's first wave of national road construction. Both have taken many decades to complete, and such completion is constantly deferred as bits of road collapse and require repair. We have illustrated how the stories of their emergence as more or less stable surfaces are ongoing stories of material drama that lack a clear point of origin or a single moment of completion. A sense of an ongoing process of material flows and blockages is echoed by the histories and accounts of the social dramas played out on the roads as they have been routinely engaged as spaces of protest and negotiation in the political life of the country. We have also traced how over the years these roads have been mobilized to channel commerce and economic potential, and how their future prospects are now threatened by new infrastructural imaginaries regarding how best to generate new national economic interests.

Yet although roads are clearly ongoing processes, what was key for the engineers with whom we worked was to find ways of delimiting their engagement with the road in order that they could demarcate the temporal and spatial boundaries of their own infrastructure project, as well as the present and future issues and relations for which they should and should not be held responsible. To do this, they deployed a range of devices and techniques that enabled them to act in anticipation of a future where they would at some point hand over the road to the nation. As we have shown, an adherence to reason, abstraction, and normative procedure appeared to offer engineers a way forward in this work, shaping the knowledge practices that have come to define their particular mode of expertise. So how might we specify the politics dynamics of such interventions?

In section 2 we have described the instruments and practices that engineers deploy to both control and keep their distance. In particular, we looked at the laboratories where soil samples are tested and manipulated; at the articulation and imposition of codes of conduct directed at the regulation of working practices; and at the transparency measures designed to combat the endemic corruption associated with public works. Each of these fields of practice mobilizes the familiar instruments of engineering practice—the standards and norms of professional codes and legal provision; the reliance on metrics, algorithms, and computer models to convert concrete measures into projected trends; and the use of technological devices to extend and standardize human perception and analytical capacity. There is no doubt that civil engineering practice enacts the framings of modern disciplinary knowledge.[8] And yet our ethnographic observations also suggest that this is a story of limited control in which the engineers' disciplinary moves are ultimately only ever going to be provisional attempts to suppress the proliferation of alternative framings. It is at this point that we find that ethnographic understandings of engineering expertise begin to reconfigure our understandings of the political.

Both Foucault (1969, 1971, 2007) and Mitchell (2002) have provided fascinating diagnostic accounts of political economy, but to do so they have had to step back from the messiness of practice. Their accounts of disciplinary power enact a scalar shift, or a move away from the complexity of the particular to the strong general account of epistemes, epochs, and regimes. As ethnographers we appreciate the clarity that such distance affords, but at the same time we remain intrigued by the politics of the specific and interested in how the space of the political might be understood from this perspective. When we attend to what civil engineers are actually doing on a day-by-day basis, we are faced with the inherently pragmatic and flexible nature of these practices. Thus, for example, we find that while standardizing metrics and the universals of mathematics are central to engineering practice, the measurements these techniques afford are never assumed to be stable. The calculations of optimum material interventions are made under laboratory conditions where materials are approached as if conditions inside and outside the laboratory were continuous—but in the knowledge that such continuity is only approximate, and that good solutions aim to be satisfactory rather than correct in any absolute sense. Both measurement and experimentation are ongoing; while experience guides the expert engineer, they know there are no ready-made solutions for the material challenges they face. Similarly, in the field of health and safety we found that stringent rules and regulations were drawn up, communicated, and enforced in the sure knowledge that most people were sidestepping and bending the rules most of the time— and that in many ways such flexibility was necessary to safe working. The regulations thus refer to a utopian world where the application of logic and the adherence to rules assumes stability and ensures that things will go according to plan. But in practice, of course, things go wrong all the time, the rules frequently contradict common-sense understandings, and the aspirations to certainty and stability run counter to the basic need for everyone to be attentive to the dynamics of the relations in which they are immersed, ready to deal with the unexpected. Finally, with respect to corruption, the regulatory instruments and legal structures designed to ensure transparency do little to assuage the sense that opaque and undisclosed relations and arrangements guide the core decisions in construction projects. Public information can always conceal the details of private negotiation.

Our interest in the specifics of how boundary-making devices are deployed in practice thus leads us to a very different, and we would argue more nuanced, understanding of the political that recovers and repositions the figure of the engineer. James Scott's critique of high modernism in *Seeing Like a State* exemplifies the broad political framing that traces the destructive failures of specific attempts at social engineering to the hubris of planners and engineers (Scott 1998). The image of the engineer as a detached, autonomous, rationalist planner is habitually produced, perhaps particularly in social anthropology, in contradistinction to alternatives, often celebrated for their more, emergent, or processual engagement with the world. Lévi-Strauss (1966) defines the engineer as the prototypical modern thinker, in contrast with the *bricoleur*, his nonmodern counterpart. In a similar vein, via a very different philosophical route, Ingold (2011) evokes the rigidity of the engineer setting out to get from point A to point B, in contrast to the "dwelling" or "wayfaring" modes of engagement that he favors as more authentic ways of living. Our argument is that while these differentiated modes of practice are indeed significant to the world of civil engineers and their capacity to produce an infrastructure project as having internal coherence and an identifiable endpoint, it is in their combination that technical knowledge gains its efficacy. Engineering projects enact their ordering power in relation to emergent environmental and social processes without embracing a dwelling perspective (Ingold 2000); they address material emergence and vibrancy without embracing animism; they engage contingency, repetition, and uncertainty without discarding a commitment to linearity, progress, and change.

The engineers we worked with were not in a position to ignore the details of local engagement. They were required to attend to other knowledges, practices, and possibilities on a daily basis. And their projects rarely simply fail or succeed, for they always carry multiple intentions and possibilities, and they always change things in unexpected ways. Contrary to the stereotype, we found that engineers know perfectly well that the data they work with is provisional, and they know that they can only transform the environment by working with what is already there. In this sense we want to suggest that engineers only partially enact the ordering paradigm of approaching the environment as an external world of nature to be grasped and controlled, despite the fact their expertise has commonly been described in this way.

The argument that we make, and which we outline in more detail in the following section, is that the professional expertise of road construction engineers lies in their ability to produce resilient structures out of the dynamic relational properties of the material and social worlds in which they find themselves. To achieve this they concentrate their effort on coordinating material relations in full awareness that these relations are realized in dynamic engagement with social worlds. Aware of environmental relations and of the productivity of difference, of systemic interconnection and oscillating variation, they attend to the possibilities of framing, of provisional decontextualization that allows relations to be stabilized long enough for decisions to be taken and actions performed. Indeed, the central aspect of their work is the capacity to frame their own expertise as a particular quality of knowledge that enables controlled and ordered material transformation as a primary responsibility undertaken, secondarily, with regard to the social implications of their work. This ability to carve out a specified domain of expertise is key, we suggest, to appreciating both how social responsibility appears as a problematic issue for modernist projects of transformation and how it is that the same expertise offers the best means available for responding to such problems. The point that we want to stress here is that the determination to identify and produce such framings is not founded in the distance of the person who fails to notice what is happening on the ground, but in the distance taken by the person who is all too aware that local complexity has somehow to be managed if planned transformations are to be embarked on. And it is in this respect that we began to think about engineers as both engineer and *bricoleur* in Lévi-Strauss's terms, as "recombinant scientists" working with what comes to hand to resolve the specific and localized problems that any infrastructural project produces in its articulation of the diverse encounters that constitute the grand plan or overall scheme.

Engineering the Social via Acts of Framing

How then do the civil engineers make space to act? How do they relate their space of action to the past and future manifestations of the infrastructures with which they are working? And how do they create and deploy framing devices in these complex spaces in which multiple knowledges and understandings compete for attention? Our approach here has been to follow the relational dynamics of the knowledge forms that engineers produce in the course of their work. Engineering projects generate all kinds of documents, plans, and analytical forms, each representing particular kinds of abstraction, each enacting a particular framing. These documents are themselves social forms, relational devices that are produced to serve a particular purpose. One such purpose is the enactment of engineering expertise. Engineers, in common with the scientists that Latour and Woolgar (1979) famously described in *Laboratory Life*, produce inscriptions. In some respects these detailed specifications are understood to be the central product of engineering expertise. The civil engineers that we have worked with, in both Peru and in the United Kingdom, have stressed to us over the years that engineering and construction are two quite different things. "Engineering" is the production of the technical solution, design, or prototype that is subsequently realized in the practice of construction or manufacture. In this respect, the crucial engineering work is carried out prior to the construction phase of a project. However, even if we stick for the moment with the idea that engineering involves the production of a design, we find in practice that such designs are produced in a variety of social framings that subtly shift the referent of the design. A feasibility study for projects such as those we studied in Peru will include all kinds of technical details and specifications concerning the road surface, its foundations, its materials, its form, its routing, and its potential costs and benefits, including the calculated risks with respect to environmental and social impact. These studies are highly technical, but they serve a particular purpose. They sort out and set out the relations between funders, politicians (or representatives of diverse public constituencies, or both), and the construction company. The engineering in this framing combines social and technical knowledges in a specific way. The studies have to address questions such as what kind of road do we want and expect this to be? And will this project be related to other concerns and policies with respect to public investment? The studies are ultimately consolidated in the contractual agreements that are drawn up to allow a project to start through the release of funds.

Quite different knowledges are assembled in the laboratories of the construction companies once the work is under way. Here, the key relations to set out and sort out are those between the available materials and the environmental forces that they expect to confront. The engineers measure and model the relational capacities of the materials with respect to things like their relative resistance to weight, plasticity, or porosity; they gauge the relative value of natural over man-made materials; and they work to find the best fit between what they have to work with and the agreed specification of their final product. These are still activities that move toward the production of an engineering design. The temporality has shifted, however. The design is still prior to the subsequent act of construction, but it is also recursive (Kelty 2005).

The problems and challenges the construction process produces are continually referred back to the laboratory for modification and subsequent design refinement.

The notion that engineering work is always prior to and separate from construction is thus itself a somewhat abstract or ideal account of how construction proceeds in practice. Engineering design informs the construction process and signals what it is that has to be built and how, but the design remains open to modification. The technical studies direct the proceedings, but they do not in the end determine how to proceed. The realization of the design is something that is worked out on the ground in the interactions between engineers (contractors and supervisors), foremen, laborers, materials, and machines. The importance of the technical specifications lies in their capacity to clear the way for action by setting out the parameters of the material transformations that are to be undertaken. However, we suggest that they also serve another, equally important function in the ways that they simultaneously delimit the relational domain for which the engineers are responsible. In this respect, the framings entailed in the drawing up of a technical specification are not predicated on the failure to address local conditions. They can perhaps be seen as attempts precisely to localize the space of intervention, to articulate its specificity, and to limit responsibility for all that will, inevitably, overflow this space at some unspecified future date. In this respect, technical knowledge works by closing down alternatives.

This formulation would, we imagine, be contested by many engineers, who often articulated their point of departure as the specification of alternatives. These alternatives should always be explored at the stage of the feasibility study to allow those in a position to decide what solution best addresses what problem. But this is not our point. There are many ways to build a road, but to the extent that any of these ways need to pass through the devices of the feasibility study, the engineering specification, the laboratory analysis, and so on, they are subject to framings that create a meaningful distance between a sphere of expert practice and the everyday worlds of the surface dramas described earlier.

These acts of framing produce social problems, in fact "frame" the social as a problem. They do so not primarily because they distance the expert from local relations, nor because they cast the expert as governing subject able to impose a particular (erroneous) truth about reality, but because their particular way of engaging the local generates a space of externality that appears, in retrospect, as problematic precisely because it is discontinuous and nonintegrated (Callon 1998). In the concern that technical projects should integrate the social, the "social" appears, by default, as that which the technical has failed to carry forward, a relational space that is disengaged and left behind. It is in this way that expert knowledge practices produce knowledge gaps, and in contexts where development and progress have widespread purchase as idioms of social improvement, such gaps signal both temporal and moral lag.

One of the paradoxes that engineers are constantly battling with is their awareness that the same people who campaigned tirelessly to secure investment from governments for roads, which they believe will deliver them better lives, are also working against the realization of this dream, even when the road is under construction and requires only their cooperation for successful completion. Such dilemmas were vividly communicated to us by one of the engineers who had worked on the Iquitos-Nauta road. He told us, admittedly with a marked note of irony in his voice, that the road workers and people living along the road that he was trying to complete were total savages. He explained that this project was the very first in Peru to use a cutting-edge synthetic webbing that had been found to be highly effective in strengthening the road surface in sites (such as the Brazilian Amazon) where there was no stone. He and his colleagues had built a shelter to keep the webbing dry and had posted a guard to keep it safe, but the guard fell asleep and the webbing was stolen. The next thing he knew, the webbing started appearing as lining for the pens in the chicken farms along the road. To make matters worse, when he finally started to use the webbing for the job it was intended for, his workers had managed to destroy it. A special machine had been brought in from Brazil to help lay the webbing on the ground, but the driver had not realized that the grooved wheels, which facilitated the laying of the material in forward drive, would destroy it in reverse. The engineer's complaint focused on how local people did not appreciate the value of the webbing, nor did they seem to appreciate that they were damaging the very thing that they were supposed to be creating together. These failures to integrate people into a common project, as framed by the civil engineers, are widely seen as the kinds of "social" issues that the specialized social relations departments of engineering companies are charged with handling.

The social relations officers are sent out to clear the way for the construction process, negotiating the minute details of routing, the sourcing of local materials, and the negotiation of compensation and work opportunities. In the process they find themselves engaged in discussions of incommensurable values and habits that are hard to reduce to the frames of developmentalist thinking, whereas infrastructural provision assumes smooth flow and prioritizes connectivity as the precondition for effective social aggregation. But the dilemmas that the social relations officers face disrupt these assumptions. How, for example, do you compensate a family for a precarious, illegal roadside shack that is their only shelter and source of livelihood? How do you negotiate the value of land that has been painstakingly tended over many decades despite its poor productivity? How do you relocate a powerful shrine that indexes forces of regeneration and future possibility for those who attend to the powers of miraculous Christian saints? How indeed do you convince people that your inadequately guarded and clearly valuable material should not be taken and used to meet a spontaneous or pressing need? Again, the social problems emerge as quite external to the technical framings, and they are associated with a nontechnical, recalcitrant, or ignorant people who do not understand how their actions impede the progress they have campaigned for. The possibility that such people might understand this perfectly and nevertheless choose to act otherwise tends not to be examined. Integrating the social in the final analysis is about getting people on board and signed up to a common project.

As others have pointed out with respect to the contemporary discussion of participatory methods in development projects (Green 2010), technical framings are thus not primarily knowledge claims; they are attempts to negotiate the ways in which a space of action (the construction project) is cut off from the everyday worlds of material and social flux and uncertainty. As we have seen, in the context of endless disputes about the origins, boundaries, and future trajectories of any infrastructural form, it is essential for engineers to produce precise framings as to what aspect of the infrastructure they are expected to be responsible for. Where the framings are put under pressure by competing agendas, alternative priorities and values emerge as social problems. The problems might be articulated as either a failure to embrace the values of progress or a failure to recognize risk. In either case education, or at least some kind of awareness raising, is nearly always produced as the answer, the means by which the integration of the social and the technical can be achieved or restored. In the process, social responsibility remains securely bracketed off from the activities of the technical expert.

It is perhaps for this reason that large-scale public works are widely thought (even by engineers themselves) to be steeped in corruption. With no secure identification of social responsibility, corruption stories flow freely around these projects. The nebulous force of interests, largely unspecified, is always thought to lie behind the decisions of who to employ, where to route a road, what materials to use, where to source them from, or where to dispose of them. Fears that the wrong kinds of integration are being effected are common-place, and the question of who gets drawn into the projects and how occu-pies everybody's thoughts. Even spaces without roads can be configured in this paradigm as a space where a road might have been had somebody not run off with the money.

As we have argued throughout this book, the skills of the engineer do not simply lie in the capacity to produce abstractions from the messy details of real world material and social processes. They also have to produce ab-stractions that are generative of possibilities for subsequent intervention in and transformation of these material and social processes. Such is the speci-ficity of engineering as opposed to the theoretical sciences. Engineering knows itself as science in action. The specificity of engineering practice is thus not about an incapacity or refusal to engage the oscillations and instabilities of the world. The engineers we worked with did, however, bring a particular orientation to this engagement, which is the specific analytic determination (supported by the instruments and institutions of modern science and tech-nology) to separate things out and to detach themselves and their focus of attention from the ongoing relational currents in which we all live. In this respect, the knowledges that these engineers produce do have a particular quality, namely the capacity of the specific abstraction to manifest as generic (i.e., as nonspecific) and thus float free of the complex relationality of the so-cial.[9] We would suggest, furthermore, that all expert knowledges rest at some point on this ability to generate authoritative generalizations from an atten-tive analysis of specific relations. Our interest in infrastructural formations is an interest in how these movements between the general and the specific are enacted, registered, and authorized. The engineers we followed in the course of our ethnography have shown us how diverse modes of integration coexist, how scalar shifts occur, and how uncertainties are addressed. Most important, perhaps, they have shown us how a focus on material transfor-mation opens up a world of tensions, negotiations, and contestations that ex-tend our understanding of political life.

Notes

Preface

1. Throughout the text we use Ing. as the common abbreviation for "ingeniero" the Spanish term for engineer, commonly used in Peru as a term of address for those with professional technical expertise.

2. CONIRSA S.A. is the registered trading name of the construction consortium created in 2005 to construct and maintain sections 2 and 3 of the Interoceanic Highway in Peru. Directly associated with IIRSA (Iniciativa para la interacion de la infraestructura regional suramericana - Initiative for the Integration of the Regional Infrastructure of South America), CONIRSA is a joint venture of Odebrecht (70%), Graña y Montero (19%), JJC Contratista Generales (7%), and ICCGSA (4%).

Introduction

1. The translation is from Lévi-Strauss (1966).

2. Examples include Boellstorff 2008; Collier 2011; Fischer 2007; Gupta 2012; Hayden 2003; Ho 2009; Latour and Woolgar 1979; Marcus 1995; Maurer 2005; Ong and Collier 2005; Riles 2001, 2011.

3. In a recent article, Marshall Sahlins (2010) recalls the occasion when he presented a paper on Pacific exchange systems at Lévi-Strauss's weekly seminar, declaring that, in contrast to his intellectual host, he was interested in infrastructures rather than the superstructures to which Lévi-Strauss had turned his theoretical attention. Lévi-Strauss had responded not only that structuralism should be extended to address the infrastructural but that conceptual structures and material practices were mutually constitutive. Good anthropology, according to Lévi-Strauss, involved holding these two possibilities as continuous rather than discontinuous possibilities. "Rather than a discontinuity, temporal as well as ontological, wherein culture appears as the symbolic afterthought of a material practice that has its own rationality, what is entailed in infrastructuralism is the realization of encompassing conceptual schemes in the particular material function of provisioning the society" (Sahlins 2010, 374–75). By contrast to the argument of Sahlins and of Lévi-Strauss, we are interested in exploring the social dynamics of material relations alongside the material dynamics of social relations.

4. A reflexive analysis of Western forms of knowledge production has been key to this critique. Whereas anthropologists have themselves been highly reflexive about their own ways of knowing, other key references in these debates are Foucault (1971) and Latour (1993).

5. For a beautiful ethnographic elaboration of this theme, see Kernaghan 2012.

6. See Hildyard 2012.

7. This latter concern is particularly evident in the work of Marilyn Strathern (1998, 2000a, 2000b, 2000–2001, 2004) and many of her students (Edwards 1993; Riles 2001), and it has been hugely influential in forging the possibility of an open and receptive conversation between science studies and ethnographic work. Recent influential work on the power of "things" includes Bennett 2009; Bennett and Joyce 2010; and Henare, Holbraad, and Wastell 2006.

8. See Larkin (2013) for a useful review of anthropological approaches to infrastructures. Other key works include Bear 2007; Chalfin 2010; Jensen and Winthereik 2013; Larkin 2008; and the Lockrem and Lugo curated collection of *Cultural Anthropology* (2012). Key interventions on Latin American include Campbell and Hetherington 2014 and Carse 2014.

9. See Graham and Thrift 2007.

10. See, for example, the classic study by James Scott (1998) and, in a different vein, the ethnographic work of Caroline Humphrey (2005).

11. Tsing (1993) has discussed the failure of the promise of infrastructures to deliver in Indonesia; and Navaro-Yashin (2012) has explored the relationship between infrastructures and social differentiation in Cyprus. Within the history of science, key works on the politics of infrastructural systems include Bowker and Star 1999; Hughes 1983, 1989; Lampland and Star 2009; Mukerji 2009; Summerton 1994; Thévenot 2002; and Winner 1986.

12. See Latour 1996.

13. See Merriman (2005), Moran (2009), and Sinclair (2002) for recent studies focusing on the intersection between the imaginative and material lives of roads in the United Kingdom.

14. For a normative account of the relationship between infrastructural capacity and "successful" national projects, see Michael Mann's (2002) discussion of infrastructural power.

15. See Joyce 2003.

16. Barry 2001; Castells 1996; Graham and Marvin 2001; Joyce 2003; Mitchell 2002; Mukerji 1997, 2009.

17. See Mitchell 2002 and the work of Foucault more generally.

18. This image of the engineer is made explicit in Lévi-Strauss's famous passage in *The Savage Mind*, where he juxtaposes the concrete recombinant science of the *bricoleur* with the abstract conceptual approach of the engineer (Lévi-Strauss 1962). In a different theoretical tradition, Ingold (2007) also exploits the contrast between modernist engineering (as a key realm of abstract reasoning) and the processual environmental engagement that characterizes the embodied skills of the craft practitioner. We return to these arguments in the conclusion.

19. We are particularly indebted to the insights of Carroll 2006; Joyce 2003; and Mukerji 1997.

20. Carroll (2006) explores the concept of engine science in his exploration of the relationship between science and government in colonial Ireland between the seventeenth and the nineteenth centuries.

21. See Burnett 2000; Edney 1999; and Prakash 1999.

22. See Mosse 2004.

23. See Ash 2000; Carroll 2006; and Mukerji 1997, 2009.

24. The reflexive dimensions of contemporary knowledge have been drawn out in helpful ways through ethnographic research by, among others, Corsín Jiménez 2007; Law 2004a and 2004b; Maurer 2005; Riles 2001; and Strathern 1996.

25. Examples include Bowker and Star 1999; Carroll 2006; Gunn 2002; Henderson 1998; Latour and Woolgar 1979; Law 2002a and 2002b; Mukerji 2009; and Suchman 1987.

26. This way of accommodating the social and technical characterized the positions of even the most outspoken critics of the sociology of science in the so-called science wars of the 1990s and 2000s. See, for example, Bricmont and Sokal (2001) for an account that stops short of the recognition of those social dynamics integral to "matters of fact" (Latour 2004).

27. Also see Trentmann's (2006) discussion of how the combination of a focus on relationality and materiality constitute a significant shift from established Foucauldian frameworks, characterized by an attention to regimes of control and a focus on how materials, concepts, and persons are interconnected. On the subject of the way in which different and often incompatible aspirations become part of road construction projects, see Thévenot (2002). Thévenot's analysis of the controversies arising around a French road construction project highlights how, in practice, modern infrastructural systems both integrate and divide, actively producing the modern territorial state as a space in which identity and discontinuity are constantly held in tension.

28. Dimitris Dalakoglou (2009) has noted that contemporary ethnographies of road construction disrupt the classic ethnographic tendency to study communities that were by definition "off road." In contrast to studies where ethnographers were concerned to

describe ways of life that were not predicated on the daily interactions with the forces of modernity that roads implied, as roads began to appear on the ethnographic scene they came accompanied by an interest in the intersections, crossovers, mobilities, and mixtures that reconfigured hegemonic understandings of modernity. Recently, a number of ethnographers, including Dalakoglou, Jeremy Campbell, Gabriel Klaegar, Daniel Normark, and Edward Simpson, have begun to explore the potential that roads offer as research sites and as intellectual objects, particularly in relation to the ways in which they encourage thinking across established domains of scholarly reflection in an attempt to show how global forces and mundane relations coexist and shape each other (see, for example, Dalakoglou and Harvey 2012).

29. For further elaboration of our research methods, see Harvey and Knox 2011.

1. Historical Futures

1. Wilson 2004.

2. Harvey 2012a; Kernaghan 2012.

3. See Christie 2008; Dean 2007; Hyslop 1984. See also http://whc.unesco.org/en /qhapaqnan/ for a discussion of a UNESCO project to nominate the Inka road network, known as the Qhapaq Ñan, as a World Heritage site.

4. For contemporary discussion of the politics surrounding this project, see Llosa (2003) and Paredes Pando (1992). One of the fears of those Limeños who make a living from the tourist industry is that this road will enable the development of tourist circuits between Brazil and Cusco that will totally sideline Lima.

5. http://bruce.graham.free.fr/family/santo_domingo/other_refs.htm.

6. See Garcia Morcillo 1982.

7. See Taussig 1987.

8. Werner Herzog's 1982 film *Fitzcarraldo* dramatically portrays the life of this rubber baron and his crazed determination to find and transport his ship across the narrow strip of dividing the waters of the Madre de Dios from those of the Marañon.

9. See Santos Granero and Barclay, 2000.

10. This was the Ley de Conscripción Vial o Servicio de Caminos (No. 4113).

11. See Nugent 1997, 177. It is interesting to note the direct parallels to the ways in which Mukerji (2010) recounts how Louis XIV attacked the enclave of the French nobility in very similar ways through the construction of the canal du Midi.

12. http://bruce.graham.free.fr/family/santo_domingo/other_refs.htm.

13. Sistema Nacional de Apoyo a la Movilización Social, the National System for Support of Social Mobilization.

14. Peru's Agrarian Reform was undertaken by a radical military dictatorship, under the leadership of General Juan Velasco Alvarado. The focus was the modernization of the agricultural sector and the redistribution of land to the rural poor. Despite the widespread expropriation of land by the state, it is widely acknowledged that the reforms did little to solve the unequal distribution of land. However, it did have the effect of restructuring and modernizing the agrarian sector, through the widespread introduction

of cooperatives under the technical guidance of SINAMOS (which also translates as without masters, *sin amos*).

15. See Kemp 2002.

16. The sol-oro was introduced as the national monetary unit in 1930, replacing the Peruvian pound.

17. Llosa 2003.

18. Between 1964 and 1980 the population of Iquitos increased from 76,000 to 165,000 (Rodriguez 1991).

19. See Wahl, Limachi, and Barletti 2003.

20. Ibid.

2. Integration and Difference

1. There was considerable press interest in the so-called *no-contactados* (uncontacted people) during the construction phase of the Interoceanic Highway. Helicopters scanned the forest for images of these people in what appeared as a response to the contradictory compulsions to prove that they existed, to demonstrate the threat the road posed to them, and to stimulate calls for their protection or integration, or both.

2. See Lopez Parodi 1991.

3. This was particularly striking as it was taking place as the same time as a broader process of land reform that dismantled the hacienda system of land ownership in the Andes, redistributing the land from the hacienda owners to communities (Seligman 1995).

4. Anon. 1989.

5. At the same time, from the perspective of many people living in Iquitos and Nauta, the road still stands as a place that is characterized by its lack of sociality. The agricultural associations along the road are often accused of being both impoverished and conflict ridden, with a number of charities and NGOs having come to focus on the road in recent years as a space in particular need of economic, educational, medical, and social intervention.

6. See Rodríguez 1991.

7. After we had completed our fieldwork on the Iquitos-Nauta Road we had the opportunity to talk with Françoise Barbira-Freedman and Stephen Hugh-Jones, who were working to set up the Ampika natural health-care project with Peruvian traditional healers. The Ampika "Live Pharmacy" (http://www.ampika.co.uk/index.html), which is located on the Iquitos-Nauta Road, is run by indigenous people who moved to the area to work and to cultivate and distribute medicinal plants.

8. See Regan 1993 and Gow 2007.

9. Programa de Formación de Maestros Bilingües de la Amazonía Peruana (FORM-ABIAP).

10. As Friere (2003) explains in more detail, environmental concerns and indigenous land rights claims in Latin America have become increasingly closely aligned.

11. See Dourojeanni 1996.

12. Ministerio de Fomento y Obras Publicas 1963.

13. For a fuller description of relationship between riches and danger in the gold economy in Latin America, see Taussig (2004).

14. The locations of other engineering camps built as part of this project.

15. A mix of Portuguese and Spanish (*portugués* and *español*), referring to the Brazilian and Peruvian engineers.

16. A mix of Quechua, the indigenous language spoken in this part of the Andes, and Spanish (*español*).

3. Figures in the Soil

1. The new highway was also devastated in this same area in January 2014 by earth tremors that simply folded the smooth new surface.

2. There are numerous references that could be given here. See, for example, Allen 1988; De la Cadena 2010; Harris 1982; Harvey 1994, 2001; Gose 1986, 1994.

3. See Rotman 1987; Lave 1988; Urton and Nina Llanos 1997; Rotman 2000; Verran 2001; Maurer 2005; and the special issue of *Anthropological Theory* (2010) on "Number as Inventive Frontier" (Guyer et al. 2010).

4. Fabian Muniesa (2014) offers a wonderfully clear account of contemporary scholarship on performativity and sets out a particularly neat summary of Latour's (1999) argument that constructivism or performativity enacts or realizes "the real." He does so in an idiom that works well for our discussions of engineering and infrastructural transformation: "Reality, like a bridge, is constructed through a laborious process of material assemblage." The fact that it is constructed does not make is unreal. The point is that the performativity approach demands attention to how that reality emerges and is sustained. As Muniesa says of the bridge, "It does not just stand there placidly without taking the trouble to happen."

5. Diario de Debates No. 8 (Matinal), April 29, 2004, Congreso de la República del Perú.

6. Ministerio de Transportes y Comunicaciones, Estudio de Prefactibilidad de Análisis de Alternativas para la Interconexión Vial Iñapari–Puerto Marítimo del Sur (Lima: Consorcio BWAS-BADALLSA, 2003).

7. Ministerio de Transportes y Comunicaciones, *Estudio de Factibilidad de la Interconexion Vial Iñapari–Puerto Maritimo del Sur* (Lima: Ministerio de Transportes y Comunicaciones Proyecto Especial de Infraestructura de Transporte Nacional, 2004).

8. See Bonifaz, Urrunaga, and Vásquez 2001; and Lacina 2009.

9. See Dunn 2005.

10. See Tsing 2005a; and Lampland and Star 2009.

11. See Vaihinger and Ogden 1924.

12. The liquid limit is defined as the moisture content at which soil begins to behave as a liquid material and begins to flow.

13. In this respect we might say that the numbers the engineers were producing in the course of road construction were multiple, in the sense that Verran (2001) suggests. Rather than seeing numbers as universals to which different kinds of meanings can be attached in different contexts, she suggests it is more helpful to think of numbers them-

selves as multiple and different in different sociomaterial configurations. Following the work of Mol (2003), Verran argues that by approaching numbers as ontological multiplicities, we can escape the notion that differences in number use can be explained with recourse to analytic divisions between divergent cultural and historical traditions. She suggests that material practices of numbering provide us with a site for observing the cultural politics through which claims to similarity and difference emerge.

14. The plastic limit is defined as the moisture content at which soil begins to behave as a plastic material. The parameters of both the liquid and plastic limit measures are set in accordance with international standards. These standards are derived from the work of Swedish chemist Albert Atterberg (1846–1916), who devised these basic measures for analyzing the water content of fine-grained soil and the subsequent classification of soil particles.

15. Jean Lave (1988) has observed, in a similar way, how in day-to-day uses of mathematics people quickly learn to find means of measuring that circumvents the need for calculation. She uses the example of Weight Watchers members to show how the weighing of food is soon replaced by alternative methods of estimating quantity, such as remembering the point on a container that corresponds to a particular volume of food or drink.

16. For a more detailed ethnographic description of gold mining in the foothills of the Andes, see Taussig 2004.

4. Health and Safety and the Politics of Safe Living

1. We draw here on a lecture delivered to the ESRC Centre for Research on Socio-Cultural Change Annual Conference, September 2013, at the School of Oriental and African Studies, University of London.

2. This very much echoes Foucault's concern with the way in which biopower operates through a concern with the modern administration of life: "a power that exerts a positive influence on life, that endeavours to administer, optimize and multiply it, subjecting it to precise controls and comprehensive regulations" (Foucault 1978, 137).

3. See Beck 1992.

4. See Stengers 2011.

5. Beckmann (2004) uses the idea of the assistant to describe those entities that enable displacement of the anticipation of harm so that action can occur. In his work, the notion of the assistant specifically describes car safety mechanisms, the webs of expertise, and the affective relations that produce the relationship of trust between the driver and the car.

6. See Harvey 2001.

7. See Allen 1988; Gose 1986, 1994; Harris 1982; Sallnow 1987.

8. See Povinelli 1995; Briggs and Mantini-Briggs 2002; Cruikshank 2005; and Kirsch 2006.

9. See Harvey 1994.

5. Corruption and Public Works

1. Stories of corruption and revelation do pertain to what Sloterdijk (1983) and Žižek (1989) have referred to as a mode of cynical reason in which a commitment to "transparency" might also entail the means through which to secure advantage. Such indirect instrumentalism is highly relevant to our argument about a relation that hinges on the dynamic tension between secrets and revelation.

2. Anthropologists have long been interested by the social entailments and effects of gossip and rumor. For an orientation to work in this field, see, for example, Gluckman 1963; Brenneis and Myers 1984; Rapport 1996; Pels 1997; Das 1998; Stewart and Strathern 2004.

3. See Strathern 2000a.

4. Gonzales Reategui was elected senator of the Center-Right Christian democratic party, the Partido Popular Cristiano, in the 1990 Fujimori government. He supported Fujimori following the auto-coup of 1992 and was rewarded with the presidency of one of the newly formed transitory councils of regional administration (CTAR), that of Loreto.

5. Dominic Boyer's discussion of conspiratorial knowledge in East German politics and history makes a similar point about how such knowledge does not necessarily work to make visible what is otherwise obscure, but rather to provide shelter, "to emancipate one's sense of self, however fleetingly, from history and identity" (Boyer 2006, 336).

6. The ethnographic work of Wendy Coxshall (2005) taught us much about this context, as did key publications by Degregori (2003, 2009, 2011).

7. CTAR is the Consejo Transitorio de Administracion Regional (Transitional Council for Regional Government) set up by President Fujimori in 1992 as part of the process of political decentralization.

8. The claim that a strong politician can be forgiven for some corrupt dealings as long as things get done is by no means unique to Peru. Lazar (2008) found exactly the same idiom at work in Bolivia, and much of the work on so-called corruption in Africa also makes this point (Comaroff and Comaroff 1999; Jordan Smith 2007; Roitman 2005).

9. http://www.transparency.org/whoweare/history.

10. For further discussion of these points and for a comparative analysis of the SNIP in relation to other transparency devices, see Harvey, Reeves, and Ruppert (2013). See also John (2011) for discussion of these issues in the context of Scottish freedom of information legislation, and Hetherington (2011) in the detailed ethnographic study of the politics of transparency in contemporary Paraguay.

11. See World Bank Resource document produced by ProInversion, Peru's Private Investment Promotion Agency: http://siteresources.worldbank.org/INTEASTASIA PACIFIC/Resources/226262-1309540769401/Session3_Godos.pdf.

12. Stephen Kass explains the sidestepping of regulation in the following way: "The financing scheme also enables the project to bypass normal review channels because, rather than being funded directly by the national governments, the Interoceanica will be paid for through revenue bonds that will, theoretically, be amortized by Interoceanica tolls. The bonds for the Peruvian portion of the Interoceanica are, however, guaranteed by the Peruvian government. This indirect financing has allowed the Interoceanica to

bypass the planning and budgetary constraints that would normally be imposed by the Peruvian government, even if, as seems highly likely, the bonds are not fully amortized by toll revenues." Online resource reprinted from the *New York Law Journal* (2009), http://www.clm.com/publication.cfm?ID=260.

13 .For a more detailed discussion of this point with reference to another road project in Peru, see Pinker 2013.

6. Impossible Publics

1. "Al final podemos darle un ultimatum. Los de supervision [the state regulator] sabe y ellos pueden intercedir."

2. Briggs and Mantini-Briggs 2003; Hayden 2003; and Jimenez 2006.

3. Ley 27117.

4. Our thanks to Dominic Boyer for this formulation, which captures the force of our argument.

5. Another ethnographer of Peru who has described this refusal to be drawn into the terms of "rational" debate is Fabiana Li (2009), who discusses how Andean protestors "step outside the document" to avoid the collaborations that public participation involves. Other examples include Kregg Hetherington's work on the protests against soya production in Paraguay (Hetherington 2013).

6. See Stengers (2005) for a discussion of the position of the "idiot" in political discourse.

7. In colloquial Spanish *"no me da la gana"* translates more easily as "I don't feel like it." However, in this case the intransitive construction suggests that what is at stake is something more than the agentive individual not being willing to engage. The phrase suggests that it was rather that this feeling did not lodge in him, and thereby move him to collaborate. Neither resisting nor complying, Aurelio is simply unmoved by Delia's request for a rational explanation.

8. McCall 1989; Deleuze 1998; Agamben 1999; and Stengers 2005.

9. See Strathern 1987, 1991, 2002.

10. By pointing to alternative modes of publicity, our aim is not to demarcate a specifically Andean mode of participation, but rather to multiply the potential modes of public participation that we might find in sites of infrastructural intervention.

11. See Sawyer 2004; Kirsch 2006; and Crook 2007.

12. Strathern 1996.

13. A fear associated with the disappearances during the conflicts of the 1980s and 1990s between the Peruvian military and the Shining Path.

7. Conclusions

1. See Das (2004) for further discussion of legibility/illegibility and De la Cadena (2010) for a detailed exposition of the possibilities of thinking in terms of ontological politics.

2. Brighenti 2010; Joyce 2003; Lefebvre 1974; and Mukerji 1997, 2009, 2010.

3. Mitchell 2002.

4. See Gluckman 1940; Frankenberg 1982; and Kapferer 2010.

5. Abélès's (1988) account of President François Mitterand's inauguration of a French railway station similarly pays very little attention to the infrastructure that was being inaugurated, suggesting that, as with our other examples, these artifacts are not foregrounded in the events that celebrate their coming into being.

6. Hildyard 2012.

7. We are indebted to Gisa Weszkalnys for her exploration of infrastructure as gesture (see Weszkalnys 2013).

8. See Blaser 2009 (445) for further discussion of "framing" in relation to Foucault's lectures at the College de France (Foucault 2007).

9. For further discussion on this point, see the introduction to Harvey and Knox (2013).

REFERENCES

Abbott, A. 1988. *The System of Professions: An Essay on the Division of Expert Labor*. Chicago: University of Chicago Press.

Abélès, M. 1988. "Modern Political Ritual." *Current Anthropology* 29 (3): 391–405.

Agamben, G. 1999. *Bartleby, or on Contingency in Potentialities*. Stanford, CA: Stanford University Press.

Akrich, M. 1992. "The De-Scription of Technical Objects." In *Shaping Technology / Building Society: Studies in Sociotechnical Change*, edited by W. Bijker and J. Law, 205–22. Cambridge, MA: MIT Press.

Allen, C. 1988. *The Hold Life Has: Coca and Cultural Identity in an Andean Community*. Washington, DC: Smithsonian Institute Press.

Allen, J. 2011. "Topological Twists: Power's Shifting Geographies." *Dialogues in Human Geography* 1 (3): 283–98.

Anand, N. 2011. "Pressure: The PoliTechnics of Water Supply in Mumbai." *Cultural Anthropology* 26 (4): 542–64.

Anon. 1989. "Carretera Iquitos-Nauta: Tumba de Milliones . . . y de Illusiones." *Kanatari* 3 (129): 6–7.

Arrospide Mejia, R. 1994. "Via Interoceanica Peruana." *Revista del Instituto de Estudios Historico Maritimos del Peru* 13: 51–82.

Ash, E. H. 2000. "'A Perfect and Absolute Work': Expertise, Authority, and the Rebuilding of Dover Harbour 1579–1583." *Technology and Culture* 41 (2): 239–68.

Barber, K. 2007. *The Anthropology of Texts, Persons and Publics: Oral and Written Culture in Africa and Beyond.* Cambridge: Cambridge University Press.

Barry, A. 2001. *Political Machines: Governing a Technological Society.* London: Athlone Press.

Barry, A., and G. Born, eds. 2013. *Interdisciplinarity: Reconfigurations of the Social and Natural Sciences.* London: Routledge.

Bear, L. 2007. *Lines of the Nation: Indian Railway Workers, Bureaucracy, and the Intimate Historical Self.* New York: Columbia University Press.

Beck, U. 1992. *Risk Society: Towards a New Modernity.* New Delhi: Sage.

Beck, U., A. Giddens, and S. Lash. 1994. *Reflexive Modernization: Politics, Tradition, and Aesthetics in the Modern Social Order.* Stanford, CA: Stanford University Press.

Beckmann, J. 2004. "Mobility and Safety." *Theory, Culture and Society* 21: 81–100.

Bennett, J. 2009. *Vibrant Matter: A Political Ecology of Things.* Durham, NC: Duke University Press.

Bennett, T., and P. Joyce, eds. 2010. *Material Powers: Cultural Studies, History and the Material Turn.* London: Routledge.

Blaser, M. 2009. "From Progress to Risk: Development, Participation, and Post-Disciplinary Techniques of Control." *Canadian Journal of Development Studies* 28 (3–4): 439–54.

Boellstorff, T. 2008. *Coming of Age in Second Life: An Anthropologist Explores the Virtually Human.* Princeton, NJ: Princeton University Press.

Bonifaz, J. L., R. Urrunaga, and J. Vásquez. 2001. *Financiamiento de la Infraestructura en El Peru: Concesion de Carreteras.* Lima: Centro de Investigacion de la Universidad del Pacifico.

Born, G. 2002. "Reflexivity and Ambivalence: Culture, Creativity and Government in the BBC." *Cultural Values* 6 (1–2): 65–90.

Bowker, G. C., and S. L. Star. 1999. *Sorting Things Out: Classification and Its Consequences.* Cambridge, MA: MIT Press.

Boyer, Dominic. 2006. "Conspiracy, History, and Therapy at a Berlin Stammtisch." *American Ethnologist* 33 (3): 327–39.

Brennan, T. 2000. *Exhausting Modernity: Grounds for a New Economy.* New York: Routledge.

Brenneis, D., and F. Myers, eds. 1984. *Dangerous Words: Language and Politics in the Pacific.* Prospect Heights, IL: Waveland Press.

Brenner, N., J. Peck, and N. Theodore. 2010. "After Neoliberalization?" *Globalizations* 7 (3): 327–45.

Bricmont, J., and A. D. Sokal. 2001. "Science and Sociology of Science: Beyond War and Peace." In *The One Culture? A Conversation about Science,* edited by J. Labinger and H. Collins, 27–47. Chicago: University of Chicago Press.

Briggs, C. L., and C. Mantini-Briggs. 2003. *Stories in the Time of Cholera: Racial Profiling during a Medical Nightmare.* Berkeley: University of California Press.

Brighenti, A. M. 2010. "On Territorology: Towards a General Science of Territory." *Theory, Culture & Society* 27 (1): 52–72.

Burnett, D. G. 2000. *Masters of All They Surveyed: Exploration, Geography, and a British El Dorado.* Chicago: University of Chicago Press.

Cache, B. 1995. *Earth Moves: The Furnishing of Territories*. Cambridge, MA: MIT Press.

Callon, M. 1986. "Some Elements of a Sociology of Translation: Domestication of the Scallops and the Fishermen of St Brieuc Bay." In *Power, Action and Belief: A New Sociology of Knowledge*, edited by John Law, 196–233. London: Routledge & Kegan Paul.

———. 1998. "An Essay on Framing and Overflowing: Economic Externalities Revisited by Sociology." In *The Laws of the Markets*, edited by M. Callon, 244–69. Oxford: Blackwell.

———. 2011. *Acting in an Uncertain World: An Essay on Technical Democracy*. Cambridge. MA: MIT Press.

Campbell, J., and K. Hetherington, eds. 2014. "Nature, Infrastructure and the State in Latin America." Special issue of the *Journal of Latin American and Caribbean Anthropology* 19 (2): 191–94.

Carroll, P. 2006. *Science, Culture, and Modern State Formation*. Berkeley: University of California Press.

Carse, A. 2014. *Beyond the Big Ditch: Politics, Ecology, and Infrastructure at the Panama Canal*. Cambridge, MA: MIT Press.

Castells, M. 1996. *The Rise of the Network Society*. Oxford: Blackwell.

Cerwonka, A., and L. H. Malkki. 2007. *Improvising Theory: Process and Temporality in Ethnographic Fieldwork*. Chicago: University of Chicago Press.

Chalfin, B. 2010. *Neoliberal Frontiers: An Ethnography of Sovereignty in West Africa*. Chicago: University of Chicago Press.

Christie, J. 2008. "Inka Roads, Lines, and Rock Shrines: A Discussion of the Contexts of Trail Markers." *Journal of Anthropological Research* 64 (1): 41–66.

Clifford, J. 1997. *Routes: Travel and Translation in the Late Twentieth Century*. Cambridge, MA: Harvard University Press.

Clifford, J., and G. E. Marcus. 1986. *Writing Culture: The Poetics and Politics of Ethnography: A School of American Research Advanced Seminar*. Berkeley: University of California Press.

Cochoy, F. 1998. "Another Discipline for the Market Economy: Marketing as a Performative Knowledge and Know-How for Capitalism." In *The Laws of the Markets*, edited by M. Callon, 194–221. Oxford: Blackwell.

Collier, S. J. 2011. *Post-Soviet Social: Neoliberalism, Social Modernity, Biopolitics*. Princeton, NJ: Princeton University Press.

Collier, S., and A. Lakoff. 2008. "The Vulnerability of Vital Systems: How Critical Infrastructure became a Security Problem." In *Securing 'The Homeland': Critical Infrastructure, Risk and (in)Security*, edited by M. Dunn Cavelty and K. Søby Kristensen, 17–39. London: Routledge.

Collier, S. J., and A. Ong. 2003. "Oikos/Anthropos: Rationality, Technology, Infrastructure." *Current Anthropology* 44 (3): 421–26.

Collins, J. 2001. "Selling the Market—Educational Standards, Discourse and Social Inequality." *Critique of Anthropology* 21 (2): 143–63.

Comaroff, J., and J. L. Comaroff. 1993. *Modernity and Its Malcontents: Ritual and Power in Postcolonial Africa*. Chicago: University of Chicago Press.

Comaroff, J. L., and J. Comaroff, eds. 1999. *Civil Society and the Political Imagination in Africa: Critical Perspectives*. Chicago: University of Chicago Press.

Coronil, F. 1997. *The Magical State: Nature, Money, and Modernity in Venezuela*. Chicago: University of Chicago Press.

Corsín Jiménez, A. 2006. *Economy and Aesthetic of Public Knowledge*. CRESC Working Paper no. 26. Manchester: CRESC.

Corsín Jiménez, A. 2007. "Industry Going Public: Rethinking Knowledge and Administration." In *Anthropology and Science: Epistemologies in Practice*, edited by J. Edwards, P. Harvey, and P. Wade, 39–57. Oxford: Berg.

Coxshall, W. 2005. "From the Peruvian Reconciliation Commission to Ethnography: Narrative, Relatedness, and Silence." *PoLAR: Political and Legal Anthropology Review* 28 (2): 203–22.

Crook, T. 2007. "Machine-Thinking: Changing Social and Bodily Divisions around the Ok Tedi Mining Project." In *Embodying Modernity and Postmodernity: Ritual, Praxis, and Social Change in Melanesia*, edited by S. Bamford, 69–104. Durham, NC: Carolina Academic Press.

Cruikshank, J. 2005. *Do Glaciers Listen? Local Knowledge, Colonial Encounters, and Social Imagination*. Seattle: University of Washington Press.

Dalakoglou, D. 2009. "An Anthropology of the Road." PhD thesis, University of London.

Dalakoglou, D., and P. Harvey. 2012. "Roads and Anthropology." Special issue, *Mobilities* 7 (4): 521–36.

Das, V. 1998. "Wittgenstein and Anthropology." *Annual Review of Anthropology* 27: 171–95.

———. 2004. "The Signature of the State: The Paradox of Illegibility." In *Anthropology in the Margins of the State*, edited by V. Das and D. Poole, 225–52. Santa Fe, NM: School of American Research Press.

———. *Life and Words: Violence and the Descent into the Ordinary*. Berkeley: University of California Press.

De Boeck, F. 2012. "Infrastructure: Commentary from Filip De Boeck." Curated Collections, Cultural Anthropology Online, November 26, 2012, http://www.culanth.org/curated_collections/11-infrastructure/discussions/7-infrastructure-commentary-from-filip-de-boeck.

Dean, C. 2007. "The Inka Married the Earth: Integrated Outcrops and the Making of Place." *Art Bulletin* 89 (3): 502–18.

Degregori, C. I. 2003. *Jamás tan cerca arremetió lo lejos: Memoria y violencia política en el Perú*. Lima: Instituto de Estudios Peruanos/Social Science Research Council.

———. 2009. *No hay país más diverso: Compendio de antropología peruana*. Lima: Instituto de Estudios Peruanos.

———. 2011. *Qué difícil es ser Dios: El Partido Comunista del Perú-Sendero Luminoso y el conflicto armado*. Lima: Instituto de Estudios Peruanos.

De la Cadena, M. 2010. "Indigenous Cosmopolitics in the Andes: Conceptual Reflections beyond 'Politics.'" *Cultural Anthropology* 25 (2): 334–70.

Deleuze, G. 1998. *Bartleby; or, the Formula in Essays Critical and Clinical*. London: Verso.

Dostoyevsky, Fyodor. (1869) 2003. *The Idiot*. London: Granta.

Dourojeanni, M. A. 1990. *Amazonía—Qué Hacer?* Iquitos: Centro de Estudios Teológicos de la Amazonía.

Dumit, J. 2004. *Picturing Personhood: Brain Scans and Biomedical Identity*. Princeton, NJ: Princeton University Press.

Dunn, E. 2005. "Standards and Person-Making in East Central Europe." In *Global Assemblages: Technology, Politics, and Ethics as Anthropological Problems*, edited by A. Ong and S. J. Collier, 173–93. Oxford: Blackwell Publishing.

Edney, M. H. 1999. *Mapping an Empire: The Geographical Construction of British India, 1765–1843*. Chicago: University of Chicago Press.

Edwards, J. 1993. *Technologies of Procreation: Kinship in the Age of Assisted Conception*. Manchester: Manchester University Press.

Ferguson, J. 2002. "Spatializing States: Towards an Ethnography of Neoliberal Governmentality." *American Ethnologist* 29 (4): 981–1002.

Fischer, M. M. J. 2007. "Culture and Cultural Analysis as Experimental Systems." *Cultural Anthropology* 22 (1): 1–65.

Foucault, M. (1969) 1989. *The Archaeology of Knowledge*. London: Routledge.

———. 1971. *The Order of Things: An Archaeology of the Human Sciences*. New York: Pantheon Books.

———. 1978. *The History of Sexuality*. London: Penguin.

———. 2007. *Security, Territory, Population: Lectures at the Collège de France, 1977–1978*. New York: Palgrave Macmillan.

Foucault, M., and M. Senellart. 2008. *The Birth of Biopolitics: Lectures at the Collège de France, 1978–79*. Basingstoke, UK: Palgrave Macmillan.

Frankenberg, R., ed. 1982. *Custom and Conflict in British Society*. Manchester: Manchester University Press.

Fraser, N. 1992. "Rethinking the Public Sphere: A Contribution to the Critique of Actually Existing Democracy." In *Habermas and the Public Sphere*, edited by C. Calhoun, 109–42. Cambridge: MIT Press.

Friel, B. 1981. *Translations*. London: Faber.

Friere, G. 2003. "Tradition, Change and Land Rights: Land Use and Territorial Strategies amongst the Piaroa." *Critique of Anthropology* 23 (4): 349–72.

Fujimura, J. H. 1992. "Crafting Science: Standardized Packages, Boundary Objects, and 'Translation.'" In *Science as Practice and Culture*, edited by A. Pickering, 168–211. Chicago: University of Chicago Press.

Gandolfo, D. 2009. *The City at Its Limits: Taboo, Transgression, and Urban Renewal in Lima*. Chicago: University of Chicago Press.

Garcia Morcillo, J. 1982. "Nueva Conquista y Colonizacion de la Amazonia Peruana." *Historia* 16 (51): 62–64.

Gluckman, M. (1940) 2002. "'The Bridge': Analysis of a Social Situation in Zululand." Reprinted in *The Anthropology of Politics: A Reader in Ethnography, Theory, Critique*, edited by J. Vincent, 53–58. Oxford: Wiley Blackwell.

———. 1963. "Gossip and Scandal." *Current Anthropology* 4: 307–16.

Gose, P. 1986. "Sacrifice and the Commodity Form in the Andes." *Man* 21 (2): 296–310.

———. 1994. *Deathly Waters and Hungry Mountains: Agrarian Ritual and Class Formation in an Andean Town*. Toronto: University of Toronto Press.

Gourevitch, A. 2010. "Environmentalism—Long Live the Politics of Fear." *Public Culture* 22: 411–24.

Gow, P. 2007. "'Ex-Cocama': Transforming Identities in Peruvian Amazonia." In *Time and Memory in Indigenous Amazonia*, edited by C. Fausto and M. Heckenberger, 194–215. Gainsville: University Press of Florida.

Graham, S., and S. Marvin. 2001. *Splintering Urbanism: Networked Infrastructures, Technological Mobilities and the Urban Condition*. London: Routledge.

Graham, S., and N. Thrift. 2007. "Out of Order: Understanding Repair and Maintenance." *Theory, Culture and Society* 24 (3): 1–25.

Green, M. 2010. "Making Development Agents: Participation as Boundary Object in International Development." *Journal of Development Studies* 46 (7): 1240–63.

Green, S. 2005. *Notes from the Balkans: Locating Marginality and Ambiguity on the Greek-Albanian Border*. Princeton, NJ: Princeton University Press.

Green, S., P. Harvey, and H. Knox. 2005. "Scales of Place and Networks: An Ethnography of the Imperative to Connect through Information and Communications Technologies." *Current Anthropology* 46 (5): 805–26.

Greenhouse, C. J. 1996. *A Moment's Notice: Time Politics across Cultures*. Ithaca, NY: Cornell University Press.

Gunn, W. 2002. "The Social and Environmental Impact of Incorporating Computer Aided Design Technologies into an Architectural Design Process." PhD diss., University of Manchester.

Gupta, A. 1995. "Blurred Boundaries—the Discourse of Corruption, the Culture of Politics, and the Imagined State." *American Ethnologist* 22 (2): 375–402.

———. 1998. *Postcolonial Developments: Agriculture in the Making of Modern India*. Durham, NC: Duke University Press.

———. 2012. *Red Tape: Bureaucracy, Structural Violence, and Poverty in India*. Durham, NC: Duke University Press.

Gupta, A., and J. Ferguson. 1992. "Beyond 'Culture': Space, Identity, and the Politics of Difference." *Cultural Anthropology* 7 (1): 6–23.

Guyer, J., N. Khan, J. Obarrio, C. Bledsoe, J. Chu, S. Bachir Diagne, K. Hart, P. Kockelman, J. Lave, C. McLoughlin, B. Maurer, F. Neiburg, D. Nelson, C. Stafford, and H. Verran, eds. 2010. "Special Section: Number as Inventive Frontier." *Anthropological Theory* (10): 1–2.

Habermas, J. 1989. *The Structural Transformation of the Public Sphere: An Inquiry into a Category of Bourgeois Society*. Cambridge: Polity.

Haraway, D. 1988. "Situated Knowledges: The Science Question in Feminism and the Privilege of Partial Perspective." *Feminist Studies* 14 (3): 575–99.

———. 1991. *Simians, Cyborgs and Women: The Reinvention of Nature*. London: Free Association.

———. 1997. *Modest-Witness@Second-Millennium.Femaleman-Meets-Oncomouse: Feminism and Technoscience*. New York: Routledge.

Harris, O. 1982. "The Dead and the Devils among the Bolivian Laymi." In *Death and the Regeneration of Life*, edited by M. Bloch and J. P. Parry, 45–73. Cambridge: Cambridge University Press.

Harvey, P. 1994. "Gender, Community and Confrontation: Power Relations and Drunkenness in Ocongate." In *Gender, Drink and Drugs*, edited by M. McDonald, 209–33. Oxford: Berg.

———. 2001. "Landscape and Commerce: Creating Contexts for the Exercise of Power." In *Contested Landscapes: Movement, Exile and Place*, edited by B. Bender and M. Winer, 197–210. Oxford: Berg.

———. 2005. "The Materiality of State Effects: An Ethnography of a Road." In *The Peruvian Andes in State Formation: Anthropological Perspectives*, edited by C. Krohn-Hansen and K. G. Nustad, 216–47. London: Pluto.

———. 2012a. "The Topological Quality of Infrastructural Relation: An Ethnographic Approach." *Theory, Culture and Society* 29 (July–September): 76–92.

———. 2012b. "Knowledge and Experimental Practice: A Dialogue between Anthropology and Science and Technology Studies." In *ASA Handbook of Social Anthropology*, edited by R. Fardon, 115–29. London: Sage.

Harvey, P., and H. Knox. 2008. "Otherwise Engaged: Culture, Deviance and the Quest for Connectivity through Road Construction." *Journal of Cultural Economy* 1 (1): 79–92.

———. 2011. "Ethnographies of Place: Researching the Road." In *Understanding Social Research: Thinking Creatively about Method*, edited by A. Dale and J. Mason, 107–19. London: Sage.

———. 2012. "The Enchantments of Infrastructure." *Mobilities* 7 (4): 521–36.

———. 2013. Introduction to *Objects and Materials: A Routledge Companion*, edited by P. Harvey, E. Casella, G. Evans, H. Knox, C. McLean, E. Silva, N. Thoburn, and K. Woodward, 1–17. London: Routledge.

Harvey, P., M. Reeves, and E. Ruppert. 2013. "Anticipating Failure: Transparency Devices and Their Effects." *Journal of Cultural Economy* 6 (4): 1–19.

Harvey, P., and S. Venkatesan. 2010. "Faith, Reason and the Ethic of Craftsmanship: Creating Contingently Stable Worlds." In *The Social after Gabriel Tarde: Debates and Assessments*, edited by M. Candea, 129–42. London: Routledge.

Hayden, C. 2003. *When Nature Goes Public: The Making and Unmaking of Bioprospecting in Mexico*. Princeton, NJ: Princeton University Press.

Heidegger, M. 1977. *The Question concerning Technology, and Other Essays*. New York: Harper and Row.

Henare, A., M. Holbraad, and S. Wastell, eds. 2006. *Thinking through Things: Theorising Artefacts in Ethnographic Perspective*. London: Routledge.

Henderson, K. 1991. "Flexible Sketches and Inflexible Data Bases: Visual Communication, Conscription Devices, and Boundary Objects in Design Engineering." *Science, Technology, & Human Values* 16 (4): 448–73.

———. 1998. "The Role of Material Objects in the Design Process: A Comparison of Two Design Cultures and How They Contend with Automation." *Science, Technology, & Human Values* 23 (2): 139–74.

Hetherington, K. 2011. *Guerrilla Auditors: The Politics of Transparency in Neoliberal Paraguay*. Durham, NC: Duke University Press.

———. 2013. "Beans before the Law: Knowledge Practices, Responsibility, and the Paraguayan Soy Boom." *Cultural Anthropology* 28 (1): 65–85.

Hildyard, N. 2012. "More Than Bricks and Mortar—Infrastructure as Asset Class: A Critical Look at Private Equity Infrastructure Funds." http://www.thecornerhouse.org.uk/sites/thecornerhouse.org.uk/files/Bricks%20and%20Mortar.pdf.

Ho, K. Z. 2009. *Liquidated: An Ethnography of Wall Street*. Durham, NC: Duke University Press.

Huertas Castillo, B., and A. Garcia Altamirano. 2003. *Los Pueblos Indigenas de Madre de Dios: Historia, Etnografia y Conyuntura*. Lima: International Work Group for Indigenous Affairs.

Hughes, T. P. 1983. *Networks of Power: Electrification in Western Society, 1880–1930*. Baltimore: Johns Hopkins University Press.

———. 1989. *American Genesis: A Century of Invention and Technological Enthusiasm, 1870–1970*. New York: Viking.

Humphrey, C. 2005. "Ideology in Infrastructure: Architecture and Soviet Imagination." *Journal of the Royal Anthropological Institute* 11 (1): 39–58.

Hyslop, J. 1984. *The Inka Road System*. New York: Academic Press.

Ingold, T. 2000. *The Perception of the Environment: Essays on Livelihood, Dwelling and Skill*. London: Routledge.

Ingold, T. 2007. *Lines: A Brief History*. London: Routledge.

———. 2011. *Being Alive: Essays on Movement, Knowledge and Description*. London: Routledge.

James, W., and D. Mills. 2005. *The Qualities of Time: Anthropological Approaches*. Oxford: Berg.

Jensen, C. B., and K. Rodje. 2010. *Deleuzian Intersections: Science, Technology and Anthropology*. Oxford: Berghahn Books.

Jensen, C. B., and B. R. Winthereik. 2013. *Monitoring Movements in Development Aid: Recursive Partnerships and Infrastructure*. Cambridge, MA: MIT Press.

John, G. 2011. "Freedom of Information and Transparency in Scotland: Disclosing Persons as Things, and Vice-versa." *Anthropology Today* 27 (3): 22–25.

Jordan Smith, D. 2007. *A Culture of Corruption: Everyday Deception and Popular Discontent in Nigeria*. Princeton, NJ: Princeton University Press.

Joyce, P. 2003. *The Rule of Freedom: Liberalism and the Modern City*. London: Verso.

Kapferer, B. 2010. "Introduction: In the Event—toward an Anthropology of Generic Moments." *Social Analysis* 54 (3): 1–27.

Keck, F., and A. Lakoff. 2013. "Figures of Warning." In "Sentinel Devices," special issue, *Limn* no. 3. http://limn.it/figures-of-warning/.

Kelty, C. 2005. "Geeks, Social Imaginaries, and Recursive Publics." *Cultural Anthropology* 20 (2): 185–214.

Kemp, K. 2002. *El Desarrollo de los Ferrocarriles en el Peru*. Lima: Proyecto Historia–Universidad Nacional de Ingenieria.

Kernaghan, R. 2009. *Coca's Gone: Of Might and Right in the Huallaga Post-Boom*. Stanford, CA: Stanford University Press.

———. 2012. "Furrows and Walls, or the Legal Topography of a Frontier Road in Peru." *Mobilities* 7 (4): 501–20.

Kirsch, S. 2006. *Reverse Anthropology: Indigenous Analysis of Social and Environmental Relations in New Guinea*. Stanford, CA: Stanford University Press.

Knorr-Cetina, K. 1999. *Epistemic Cultures: How the Sciences Make Knowledge*. Cambridge, MA: Harvard University Press.

Knox, H. 2013. "Real-izing the Virtual: Digital Simulation and the Politics of Future Making." In *Objects and Materials: A Routledge Companion*, edited by P. Harvey, E. Casella, G. Evans, H. Knox, C. McLean, E. Silva, N. Thoburn, and K. Woodward, 291–301. London: Routledge.

Knox, H., D. O'Doherty, T. Vurdubakis, and C. Westrup. 2012. "Enacting the Global in the Age of Enterprise Resource Planning." *Anthropology in Action* 19 (1): 32–46.

Knox, H., M. Savage, and P. Harvey. 2006. "Social Networks and the Study of Relations: Networks as Method, Metaphor and Form." *Economy and Society* 35 (1): 113–40.

Lacina, P. 2009. "Comment and Casenote: Public-Private Road Building in Latin America." *American Law and Business Review* 15 (3): 661–71.

Lakoff, A. 2005. "The Private Life of Numbers: Pharmaceutical Marketing in Post-Welfare Argentina." In *Global Assemblages*, edited by A. Ong and S. J. Collier, 194–213. Oxford: Blackwell.

Lampland, M., and S. L. Star. 2009. *Standards and Their Stories: How Quantifying, Classifying, and Formalizing Practices Shape Everyday Life*. Ithaca, NY: Cornell University Press.

Larkin, B. 2008. *Signal and Noise: Media, Infrastructure, and Urban Culture in Nigeria*. Durham, NC: Duke University Press.

——. 2013. "The Politics and Poetics of Infrastructure." *Annual Review of Anthropology* 42: 327–43.

Latour, B. 1987. *Science in Action: How to Follow Scientists and Engineers through Society*. Cambridge, MA: Harvard University Press.

——. 1993. *We Have Never Been Modern*. New York: Prentice Hall.

——. 1996. *Aramis: Or the Love of Technology*. Cambridge, MA: Harvard University Press.

——. 1997. "Trains of Thought: Piaget, Formalism and the Fifth Dimension." *Common Knowledge* 6 (3): 170–91.

——. 1999. *Pandora's Hope: Essays on the Reality of Science Studies*. Cambridge, MA: Harvard University Press.

——. 2004. "Why Has Critique Run Out of Steam? From Matters of Fact to Matters of Concern." *Critical Inquiry* 20: 225–48.

——. 2005. *Reassembling the Social: An Introduction to Actor-Network Theory*. Oxford: Oxford University Press.

——. 2009. "Perspectivism: 'Type' or 'Bomb'?" *Anthropology Today* 25 (2): 1–2.

Latour, B., and S. Woolgar. 1979. *Laboratory Life: The Social Construction of Scientific Fact*. London: Sage.

Lave, J. 1988. *Cognition in Practice: Mind, Mathematics and Culture in Everyday Life*. Cambridge: Cambridge University Press.

Law, J. 2002a. *Aircraft Stories: Decentering the Object in Technoscience*. Durham, NC: Duke University Press.

——. 2002b. "On Hidden Heterogeneities: Complexity, Formalism, and Aircraft Design." In *Complexities: Social Studies of Knowledge Practices*, edited by A. Mol and J. Law, 116–41. Durham, NC: Duke University Press.

——. 2004a. *After Method: Mess in Social Science Research*. London: Routledge.

———. 2004b. "And if the Global Were Small and Noncoherent? Method, Complexity, and the Baroque." *Environment and Planning D: Society and Space* 22 (1): 13–26.

Law, J., and J. Hassard, eds. 1999. *Actor Network Theory and After.* Oxford: Blackwell.

Lazar, S. 2008. *El Alto, Rebel City: Self and Citizenship in Andean Bolivia.* Durham, NC: Duke University Press.

Lefebvre, H. (1974) 1991. *The Production of Space.* Oxford: Blackwell.

Lévi-Strauss, C. (1962) 1966. *The Savage Mind.* Chicago: University of Chicago Press.

Li, F. 2009. "Documenting Accountability: Environmental Impact Assessment in a Peruvian Mining Project." *PoLAR: Political and Legal Anthropology Review* 32 (2): 218–36.

Llosa, E. 2003. *La Batalla por la Carreterra Interoceanica en el Sur Peruano: Localismo o Decentralismo?* Lima: Instituto de Estudios Peruanos.

Lockrem, J., and Adonia L. 2012. "Online curated collection of Cultural Anthropology." http://www.culanth.org/curated_collections/11-infrastructure.

Long, N. 1989. *Encounters at the Interface: A Perspective on Social Discontinuities in Rural Development.* Wageningen, Netherlands: Agricultural University.

Lopez Parodi, J. 1991. *Evaluacion del Proyecto Programa de Apoyo al Desarollo de los Asentamientos Humanos de aa Carretera Iquitos-Nauta.* Iquitos: Agencia Española de Cooperación Internacional (AECI) and Instituto de Investigaciones de la Amazonía Peruana (IIAP).

Lury, C., L. Parisi, and T. Terranova. 2012. "Introduction: The Becoming Topological of Culture." *Theory, Culture and Society* 29 (4–5): 3–35.

Mann, M. 1993. *The Sources of Social Power.* Cambridge: Cambridge University Press.

———. 2002. "The Crisis of the Latin American Nation-State: The Political Crisis and Internal Conflict in Colombia." Paper presented to the Conference "The Political Crisis and Internal Conflict in Colombia", April 10–13, 2002. Bogota, Colombia. www.sscnet.ucla.edu/soc/faculty/mann/colombia.pdf.

Marcus, G. E. 1995. "Ethnography in/of the World System: The Emergence of Multi-Sited Ethnography." *Annual Review of Anthropology* 24: 95–117.

Maurer, B. 2005. *Mutual Life, Limited: Islamic Banking, Alternative Currencies, Lateral Reason.* Princeton, NJ: Princeton University Press.

McCall, D. 1989. *The Silence of Bartleby.* Ithaca, NY: Cornell University Press.

Melville, H. (1853) 2007. *Bartleby the Scrivener.* London: Hesperus.

Merriman, P. 2005. "'Operation Motorway': Landscapes of Construction on England's M1 Motorway." *Journal of Historical Geography* 31: 113–33.

Mimica, J. 1988. *Intimations of Infinity: The Cultural Meanings of the Iqwaye Counting and Number Systems.* Oxford: Berg.

Ministerio de Fomento y Obras Publicas. 1963. "Direccion general de coordinacion economica y financiera 1963 Estudio de Factibilidad Economica de la Carretera Quince-Mil–Yoringo–Puerto Maldonado." Lima: Ministerio de Fomento y Obras Publicas, Direccion general de coordinacion economica y financiera, Direccion cable-grafica FOMENDIC CEF.

Mitchell, T. 2002. *Rule of Experts: Egypt, Techno-Politics, Modernity.* Berkeley: University of California Press.

———. 2009. "Carbon Democracy." *Economy and Society* 38 (3): 399–432.

Mohl, R. A. 2004. "Stop the Road: Freeway Revolts in American Cities." *Journal of Urban History* 30 (5): 674–706.

Mol, A. 2003. *The Body Multiple: Ontology in Medical Practice.* Durham, NC: Duke University Press.

Moran, J. 2009. *On Roads.* London: Profile.

Mosse, D. 2004. *Cultivating Development: An Ethnography of Aid Policy and Practice.* London: Pluto.

Mukerji, C. 1997. *Territorial Ambitions and the Gardens of Versailles.* Cambridge: Cambridge University Press.

———. 2009. *Impossible Engineering: Technology and Territoriality on the Canal du Midi.* Princeton, NJ: Princeton University Press.

———. 2010. "The Unintended State." In *Material Powers: Cultural Studies, History and the Material Turn*, edited by T. Bennett and P. Joyce, 81–101. London: Routledge.

Navaro-Yashin, Y. 2012. *The Make-Believe Space: Affective Geography in a Post-War Polity.* Durham, NC: Duke University Press.

Nepstad, D., C. Stickler, and O. Almeida. 2006. "Globalization of the Amazon Soy and Beef Industries: Opportunities for Conservation." *Conservation Biology* 20 (6): 1595–1603.

Nugent, D. 1997. *Modernity at the Edge of Empire: State, Individual, and Nation in the Northern Peruvian Andes, 1885–1935.* Stanford, CA: Stanford University Press.

Ong, A., and S. J. Collier. 2005. *Global Assemblages: Technology, Politics, and Ethics as Anthropological Problems.* Oxford: Blackwell Publishing.

Orlove, B. S. 1993. "Putting Race in Its Place: Order in Colonial and Postcolonial Peruvian Geography." *Social Research* 60 (2): 301–36.

Osborne, T. 2004. "On Mediators: Intellectuals and the Ideas Trade in the Knowledge Society." *Economy and Society* 33 (4): 430–47.

Paredes Pando, O. 1992. *Carretera Interoceanica: Integración o marginacion de la region Inka.* Cusco: Bartolome de las Casas.

Pels, P. 1997. "The Anthropology of Colonialism: Culture, History and the Emergence of Western Governmentality." *Annual Review of Anthropology* 26: 163–83.

Pickering, A. 2011. *The Cybernetic Brain: Sketches of Another Future.* Chicago: University of Chicago Press.

Pinker, A. 2013. "Papeles de Doble Cara: La politica de la documentacion en un proyecto de ingenieria publica." *Anthropologica* 30 (30): 101–22.

Poole, D. 2004. "Between Threat and Guarantee: Justice and Community in the Margins of the Peruvian State." In *Anthropology in the Margins of the State*, edited by V. Das and D. Poole, 35–66. Santa Fe, NM: School of American Research Press.

———. 2005. "Los dos cuerpos del juez: Comunidad, justicia, y corrupcion en el Peru de los neoliberales." In *Vicios Publicos: Poder y Corrupcion*, edited by O. Ugarteche, 57–80. Lima: Fondo de Cultura Económica.

Povinelli, E. 1995. "Do Rocks Listen? The Cultural Politics of Apprehending Australian Aboriginal Law." *American Anthropologist* 97 (3): 505–18.

Prakash, G. 1999. *Another Reason: Science and the Imagination of Modern India.* Princeton, NJ: Princeton University Press.

Rabinow, P. 1996. *Making PCR: A Story of Biotechnology.* Chicago: University of Chicago Press.

———. 1999. *French DNA: Trouble in Purgatory.* Chicago: University of Chicago Press.

Rapport, N. 1996. "Gossip." In *Encyclopedia of Social and Cultural Anthropology,* edited by A. Barnard and J. Spencer, 266–67. London: Routledge.

Regan, J. 1993. *Hacia la Tierra sin Mal: La Religion del Pueblo en la Amazonia.* Iquitos: Centro de Estudios Teológicos de la Amazonía (CETA).

Riles, A. 2001. *The Network Inside Out.* Ann Arbor: University of Michigan Press.

———. 2011. *Collateral Knowledge: Legal Reasoning in the Global Financial Markets.* Chicago: University of Chicago Press.

Rodríguez, M. A. 1991. "Amazonia: Indigenas, Campesinos y Proletarios." *Debates en Sociologia* 16: 125–48.

Roitman, J. 2005. *Fiscal Disobedience: An Ethnography of Economic Regulation in Central Africa.* Princeton, NJ: Princeton University Press.

Romero Sotomayor, Carlos. 1935. "Impresiones Sobre un Reconocimiento Aereo para la Elección de una ruta de carretera a través de la selva Amazónica." *Informaciones y Memorias de la Sociedad de Ingenieros del Peru* 36 (11–12): 426–30.

Rotman, B. 1987. *Signifying Nothing: The Semiotics of Zero.* New York: St. Martin's Press.

———. 2000. *Mathematics as Sign: Writing, Imagining, Counting.* Stanford, CA: Stanford University Press.

Ruppert, E. 2011. "Population Objects: Interpassive Subjects." *Sociology* 45 (2): 218–33.

Sahlins, M. 2010. "Infrastructuralism." *Critical Inquiry* 36 (3): 371–85.

Sallnow, M. J. 1987. *Pilgrims of the Andes: Regional Cults in Cusco.* Washington, DC: Smithsonian Institution Press.

Salvatore, R. 2006. "Imperial Mechanics: South America's Hemispheric Integration in the Machine Age." *American Quarterly* 58 (3): 662–91.

Sampson, S. 2005. "Integrity Warriors: Global Morality and the Anti-corruption Movement in the Balkans." In *Understanding Corruption: Anthropological Perspectives,* edited by D. Haller and C. Shore, 103–30. London: Pluto Press.

Santolalla, Nicholas R. 1952. "Impresiones sobre un Reconocimiento Aereo Para la Eleccion de una Ruta de Carretera a través de la Selva Amazónica." *Informaciones y Memorias de la Sociedad de Ingenieros del Perú* 53 (1–3): 29–33.

Santos Granero, F., and F. Barclay. 1992. *Selva Central: History, Economy, and Land Use in the Peruvian Amazon.* Washington, DC: Smithsonian Institution Press.

Sawyer, S. 2004. *Crude Chronicles: Indigenous Politics, Multinational Oil, and Neoliberalism in Ecuador.* Durham, NC: Duke University Press.

Schivelbusch, W. 1986. *The Railway Journey: The Industrialization and Perception of Time and Space in the 19th Century.* Leamington Spa, UK: Berg.

Scott, J. C. 1998. *Seeing Like a State: How Certain Schemes to Improve the Human Condition Have Failed.* New Haven, CT: Yale University Press.

Scott, J. W., and D. Keates. 2001. *Schools of Thought: Twenty-Five Years of Interpretive Social Science.* Princeton, NJ: Princeton University Press.

Seligmann, L. J. 1995. *Between Reform and Revolution: Political Struggles in the Peruvian Andes, 1969–1991.* Stanford, CA: Stanford University Press.

Shore, C., and S. Wright. 1999. "Audit Culture and Anthropology: Neo-Liberalism in British Higher Education." *Journal of the Royal Anthropological Institute* 5 (4): 557–76.

Simatovic, M. I. R., M. Glave, and G. Pastor. 2008. *Impact of the Rural Roads Program on Democracy and Citizenship in Rural Areas of Peru.* Lima: Institute of Peruvian Studies.

Sinclair, I. 2002. *London Orbital: A Walk around the M25.* London: Granta.

Slater, C. 2002. *Entangled Edens: Visions of the Amazon.* Berkeley: University of California Press.

Sloterdijk, P. 1983. *Critique of Cynical Reason.* Minneapolis: University of Minnesota Press.

——. 2009. *Terror from the Air.* Cambridge, MA: MIT Press.

Stahl, Eurico G. 1925. "La Vialidad del Peru: Conferencia Sustentada en la Sociedad Geografica de Lima." *Boletin de la Sociedad Geografica de Lima* 42 (1): 29–57.

Star, S. L., and J. R. Griesemer. 1989. "Institutional Ecology, 'Translations' and Boundary Objects: Amateurs and Professionals in Berkeley's Museum of Vertebrate Zoology, 1907–39." *Social Studies of Science* 19 (3): 387–420.

Star, S. L., and K. Ruhleder. 1996. "Steps toward an Ecology of Infrastructure: Design and Access for Large Information Spaces." *Information Systems Research* 7 (1): 111–34.

Stengers, I. 2005. "The Cosmopolitical Proposal." In *Making Things Public: Atmospheres of Democracy,* edited by B. Latour and P. Weibel, 994–1003. Cambridge, MA: MIT Press.

——. 2007. "Diderot's Egg: Divorcing Materialism from Eliminativism." *Radical Philosophy* 144 (July–August): 7–15.

——. 2011. "Including Nonhumans in Political Theory: Opening Pandora's Box?" In *Political Matter: Technoscience, Democracy, and Public Life,* edited by B. Braun and S. Whatmore, 3–33. Minneapolis: University of Minnesota Press.

Stewart, K. 1996. *A Space on the Side of the Road: Cultural Poetics in an 'Other' America.* Princeton, NJ: Princeton University Press.

——. 2007. *Ordinary Affects.* Durham, NC: Duke University Press.

Stewart, P., and A. Strathern. 2004. *Witchcraft, Sorcery, Rumors, and Gossip.* Cambridge: Cambridge University Press.

Stoller, P. 2009. *The Power of the Between: An Anthropological Odyssey.* Chicago: University of Chicago Press.

Strathern, M. 1987. "The Limits of Auto-Anthropology." In *Anthropology at Home,* edited by A. Jackson, 16–37. London: Tavistock.

——. 1991. *Partial Connections.* Savage, MD: Rowman & Littlefield.

——. 1996. "Cutting the Network." *Journal of the Royal Anthropological Institute* 2: 517–35.

——. 1998. "Innovation and Creativity: What Is New in a Centennial Year?" *Cambridge Anthropology* 20 (1–2): 2–6.

——. 2000a. "The Tyranny of Transparency." *British Educational Research Journal* 26 (3): 309–21.

——. 2000b. *Audit Cultures: Anthropological Studies in Accountability, Ethics and the Academy.* London: Routledge.

——. 2002. "Abstraction and Decontextualisation: An Anthropological Comment." In *Virtual Society? Technology, Cyberbole, Reality,* edited by S. Woolgar, 302–13. Oxford: Oxford University Press.

——. 2004. *Commons and Borderlands: Working Papers on Interdisciplinarity, Accountability and the Flow of Knowledge.* Wantage, UK: Sean Kingston.

Suchman, L. A. 1987. *Plans and Situated Actions: The Problem of Human-Machine Communication.* Cambridge: Cambridge University Press.

Summerton, J. 1994. *Changing Large Technical Systems.* Boulder, CO: Westview Press.

Taussig, M. T. 1987. *Shamanism, Colonialism, and the Wild Man: A Study in Terror and Healing.* Chicago: University of Chicago Press.

——. 1997. *The Magic of the State.* New York: Routledge.

——. 1999. *Defacement: Public Secrecy and the Labor of the Negative.* Stanford, CA: Stanford University Press.

——. 2004. *My Cocaine Museum.* Chicago: University of Chicago Press.

——. 2006. *Walter Benjamin's Grave.* Chicago: University of Chicago Press.

Thévenot, L. 2002. "Which Road to Follow? The Moral Complexity of An 'Equipped' Humanity." In *Complexities: Social Studies of Knowledge Practices*, edited by J. Law and A. Mol, 53–87. Durham, NC: Duke University Press.

Thoburn, N. 2003. *Deleuze, Marx, and Politics.* London: Routledge.

Thrift, N. 2007. *Non-Representational Theory: Space, Politics, Affect.* London: Routledge.

Traweek, S. 1988. *Beamtimes and Lifetimes—the World of High Energy Physicists.* Cambridge, MA: Harvard University Press.

Trentmann, F. 2006. *The Making of the Consumer: Knowledge, Power and Identity in the Modern World.* Oxford: Berg.

Tsing, A. 1993. *In the Realm of the Diamond Queen: Marginality in an Out-of-the-Way Place.* Princeton, NJ: Princeton University Press.

——. 1999. "Inside the Economy of Appearances." *Public Culture* 12 (1): 115–44.

——. 2000. "The Global Situation." *Cultural Anthropology* 15 (3): 327–60.

——. 2005a. *Friction: An Ethnography of Global Connection.* Princeton, NJ: Princeton University Press.

——. 2005b. "How to Make Resources in Order to Destroy Them (and Then Save Them?) on the Salvage Frontier." In *Histories of the Future*, edited by D. Rosenberg and S. F. Harding, 53–72. Durham, NC: Duke University Press.

Urton, G., and P. N. Llanos. 1997. *The Social Life of Numbers: A Quechua Ontology of Numbers and Philosophy of Arithmetic.* Austin: University of Texas Press.

Vaihinger, H., and C. K. Ogden. 1924. *The Philosophy of 'As If': A System of the Theoretical, Practical and Religious Fictions of Mankind.* New York: Harcourt Brace.

Verran, H. 2001. *Science and an African Logic.* Chicago: University of Chicago Press.

Wahl, L., L. Limachi, and J. Barletti. 2003. "Del Discurso Oficial y Caserío Rural: El Desarrollo Regional y la Carretera Iquitos-Nauta." In *Amazonía: Procesos, Demográficos, y Ambientales*, edited by C. E. Aramburú and E. Bedoya Garland, 155–82. Lima: Consorcio de Investigación Económica y Social.

Wall, D. 1999. *Earth First! And the Anti-Roads Movement: Radical Environmentalism and Comparative Social Movements.* London: Routledge.

Warner, M. 2002. "Publics and Counterpublics." *Public Culture* 14 (1): 49–90.

Welker, Marina. 2014. *Enacting the Corporation: An American Mining Firm in Post-Authoritarian Indonesia.* Berkeley: University of California Press.

Weszkalnys, Gisa. 2013. "Oil's Magic: Contestation and Materiality." In *Cultures of Energy: Power, Practices, Technologies*, edited by S. Strauss, S. Rupp, and T. Love, 267–83. Walnut Creek, CA: Left Coast Press.

Wilson, F. 2004. "Towards a Political Economy of Roads: Experiences from Peru." *Development and Change* 35 (3): 525–46.

Winner, L. 1986. "Do Artifacts Have Politics?" In his *The Whale and the Reactor: A Search for Limits in an Age of High Technology*, 19–39. Chicago: University of Chicago Press.

Žižek, S. 1989. *The Sublime Object of Ideology*. London: Verso.

INDEX

Note: Italic numbers refer to illustrations.